T0331755

Schrödinger's Mechanics

Interpretation

Other World Scientific Titles by the Author

Schrödinger's Mechanics
ISBN: 978-9971-5-0760-2

Probability and Schrödinger's Mechanics
ISBN: 978-981-238-191-0

Quantum Chemistry: A Unified Approach
Second Edition
ISBN: 978-1-84816-746-9

Schrödinger's Mechanics

Interpretation

David B Cook

University of Sheffield, UK

World Scientific

NEW JERSEY · LONDON · SINGAPORE · BEIJING · SHANGHAI · HONG KONG · TAIPEI · CHENNAI · TOKYO

Published by

World Scientific Publishing Europe Ltd.

57 Shelton Street, Covent Garden, London WC2H 9HE

Head office: 5 Toh Tuck Link, Singapore 596224

USA office: 27 Warren Street, Suite 401-402, Hackensack, NJ 07601

Library of Congress Cataloging-in-Publication Data

Names: Cook, David B. (David Branston), author.

Title: Schrödinger's mechanics : interpretation / by David B. Cook

    (University of Sheffield, UK).

Description: New Jersey : World Scientific, 2018.

Identifiers: LCCN 2017052617 | ISBN 9781786344908 (hc : alk. paper)

Subjects: LCSH: Wave mechanics. | Schrödinger equation. | Schrödinger operator. |

    Quantum theory. | Numerical analysis.

Classification: LCC QC174.26.W28 C6574 2018 | DDC 530.12/4--dc23

LC record available at https://lccn.loc.gov/2017052617

British Library Cataloguing-in-Publication Data

A catalogue record for this book is available from the British Library.

For any available supplementary material, please visit
http://www.worldscientific.com/worldscibooks/10.1142/Q0143#t=suppl

Desk Editors: Herbert Moses/Jennifer Brough/Koe Shi Ying

Typeset by Stallion Press
Email: enquiries@stallionpress.com

Printed in Singapore

Preface

There is no shortage of books about the interpretation of quantum theories; they range from the 'what do I need to know' type to fantastic and unverifiable conjectures about the nature of the universe(s), why should one get involved in such a knotty area of dispute? What I would like to do in this short account is to emphasise that the original form of Quantum Mechanics is not a series of useful formulae for interpreting experimental results nor does it require any new philosophies, it is a cogent interpreted theory of the behaviour of sub-atomic particles and has been as successful in this region as Newtons mechanics has been in the familiar world. In this context we academics all write, in our 'Mission Statements' and our 'Self-evaluations of Teaching', that we encourage, even insist on, a critical, scientific attitude to learning. But *where* is this critical attitude encouraged and what *type* of criticism is to be elicited from our students?

Of course, we encourage a critical attitude to the quality and reliability of experimental data — error analysis and all that — but this is just routine, little more than a formalisation of common sense and good housekeeping. Only very rarely is a criticism of the *theories* of science and their interpretation encouraged and if it is, this criticism is to be found in formal courses in 'History and Philosophy of Science' and not in the real everyday process of confronting *current* science in our teaching and research. One is reminded of the saying "every generation must write history anew for itself"; much lip service and very little practice. In fact, if every generation of scientists does not rework the theories which they have been taught 'for themselves' (at least in outline) then science stagnates and becomes a set of recipes; just another set of axioms to be manipulated.

This work is the outcome of many years — mostly post-retirement — of musing about quantum mechanics and its interpretation. In this, I have been mainly influenced by what I have called elsewhere Schrödinger's Mechanics (henceforth SM); the mechanics developed by Schrödinger in his astonishing paper of 1926. During these years I have come to realise more and more the absolutely central importance of the method developed in this paper. I shall want to say more about the strange fact that this paper is now largely unread and the methods used in it apparently unknown. So, this work could be described as 'old hat' in that it is almost a century since this pioneering publication; but the disagreements about the *interpretation* of SM and all of the quantum mechanics which followed is certainly not so antique, indeed it is an open sore in physics.

In previous publications I have concentrated on the non-relativistic SM while occasionally promising to have a more careful look at the problem of making quantum mechanics compatible with Special Relativity, that is to see how to introduce Lorentz invariance onto the scene. Special Relativity has played hardly any part in my professional research (quantum chemistry) and I have always had a strong suspicion that the idea that electron 'spin' somehow depends on Special Relativity is suspect. This is naturally due to my prejudice, as a chemist, against Dirac's physical interpretation of his 'pretty mathematics'[a] when applied to another phenomenon — the theory of the chemical bond as 'spin coupling' — which was a red herring in the physical interpretation of the driving force of bond formation for decades. I have addressed the matter of spin in Chapter 6.

In this work I will show that, when equipped with a realistic interpretation, the methods of Schrödinger's 1926 paper contain

---

[a]*International Journal of Theoretical Physics*, **21**(8–9), pp. 603–605.

either explicitly or implicitly all other formulations of Quantum Mechanics. And I use 'contain' deliberately; SM is not 'equivalent to' other formulations, it includes them in the sense that many 'principles' (axioms) of other methods are theorems in SM. What *is* true, however, is that SM shares some formal (mathematical) structures with other theories; but that is extremely common in mathematically articulated sciences; all that this means is that certain formal mathematical structures may be abstracted from SM. What makes SM unique is its physical interpretation and its explicit and transparent links to Classical Particle Mechanics. With the historical development of methods allegedly based on or derived from SM has come a retreat from physical interpretation in general and the very idea of submitting one's theories to independent verification is often abandoned. In criticising the standard interpretations of quantum theories I am very aware of the possible shortcomings of such work as aptly described by Kenneth Tynan's 'definition': "A critic is a man [*sic*] who knows the way but can't drive the car".

A word or two about my eccentricities: I use the word abstract as both an adjective and, more often, as a verb. Also, I have purposely tended to refer to older publications since these earlier authors were struggling to understand the meaning of all the new theories in science not simply to work from what are regarded as standard interpretations. I have used appendices and endnotes as well as the occasional footnote; footnotes are indicated by the traditional symbols (*, † etc.) and the endnotes, which tend to be longer, by numbers. I hope that the book could be read simply by reading the chapters leaving appendices and endnotes for some clarification. Some principle points are repeated or referred to from time-to-time for the reader who, like myself, tends to dip into a book without reading from the start!

I am thinking of this essay as some form of *reculer pour mieux sauter*. I can try the *reculer* part and I leave the *mieux sauter* to others.

David B. Cook

About the Author

Dr. David B. Cook is an Associate of the University of Sheffield, England. He graduated from the University of Sheffield with a B.Sc. in 1962, postgraduation (PhD) in 1966, and a D.Sc. in 1989. He served as a Postdoctoral Fellow along with Prof. Roy McWeeny at Keele University, England, in the year 1966. During his tenure from 1967 to 2003 at the University of Sheffield, he has been a Lecturer, a Senior Lecturer, and a Director of Studies. His earlier published books include *Ab Initio Valence Calculations in Chemistry* (Butterworths, 1974), *Structures and Approximations for Electrons in Molecules* (Ellis Horwood, 1978), *Schrödinger's Mechanics* (World Scientific, 1988), *Handbook of Computational Quantum Chemistry* (OUP, 1998), *Probability and Schrödinger's Mechanics* (Imperial College Press, 2002), *Handbook of Computational Quantum Chemistry (Revised Edition)* (Dover Publications, 2005), *Quantum Chemistry: A Unified Approach* (Imperial College Press, 2008), and *Quantum Chemistry: A Unified Approach (Revised Second Edition)* (Imperial College Press, 2012). He has published over 60 scientific papers in the areas of the theory and implementation of the calculation of molecular electronic structure; in three-quarters of these, he was the sole author.

Contents

Chapter 1

Aims

This work is divided into two parts: the first part (Chapters 2–5) tries, in the most cogent and positive possible terms, to summarise what Schrödinger's Mechanics is and its physical interpretation. The remaining chapters are a series of comments, from the point of view of SM, on other developments in quantum theory.

The overall aims of the work, in the first part are as follows:

(1) To place Schrödinger into the position of the culmination of the 'Great Tradition' of Analytical Mechanics, that is after Newton, Lagrange, Hamilton and Jacobi. Also adding the pioneering work of Poincaré and Kolmogorov to create an interpreted, paradox-free theory of the energetics and probability distributions of sub-atomic particles; forming a clear and comprehensible formulation of what I have called Schrödinger's Mechanics (SM).

(2) To show that other 'formulations' of Quantum Mechanics (QM) are either reducible to or abstracted from SM; none of them are capable of generating SM in the same sense that SM can generate them.

(3) The probability distributions in SM, and the state functions from which they are formed, are the same as all other probability distributions. They are mathematical functions of space and exist not in the real world but in our heads, on paper or in our computers. The relationship between Probability and Statistics is emphasised. Confusions between these two and the role of

1

'colloquial' uses of language in discussions of probability and physics can be seen as a source of most difficulties.

When this is done all the 'traditional problems' of the interpretation of QM simply fall away. There is no wave–particle duality; no measurement problem; no collapse of the wave function and, most disappointingly for popular science writers, no 'many-worlds' or 'many-universes' speculation. Heisenberg's uncertainty 'principle' is shown to be a theorem and not a universal excuse for creating what (in a different context) Stephen Jay Gould labeled 'Just-So Stories': *post hoc* justifications of current practice.

The remaining chapters address points of wider interest:

(1) To show the interpretation of SM can be extended to include 'spin' entirely within non-relativistic SM by substituting a Geometric Algebra representation of space for the usual symbolic vector system.

(2) To investigate some consequences of imposing the constraints of Special Relativity on the formulation of SM.

(3) To try and draw a sharp distinction between a physical theory and the mathematical technologies used in the approximation methods involved in practical applications; to distinguish interpretation of a theory from attempts to interpret the mathematical terms generated by these technologies.

(4) To distinguish between physical and mathematical 'fields' and the consequences of this distinction for the physical interpretation of these entities. To question the conflation of fields and particles.

(5) To stress the limitations of what can be achieved by this approach. Any experimental data which can be conclusively shown to be inexplicable by the methods outlined here call for an extension of the 'Great Tradition' beyond SM and will not be considered here. No theory is of unlimited application,[a] but any scheme which extends or replaces SM would have to satisfy the strict requirements of the tradition; it would have to be

---

[a]There will never be a 'Theory of everything'.

consistent, interpretable and, above all, at least as successful in describing the *mechanisms* of the real world and generate verifiable physical results. Finally, SM has been at least as successful in interpreting and describing the sub-atomic world (electron physics and chemistry) as Classical Particle Mechanics has been on the human scale. Any acceptable theory of sub-nuclear phenomena would be required to be as successful and not yet another example of semi-empirical 'Ptolomaic Science'.

It might be thought that the work presented here is 'old-fashioned' and, indeed, so it is, but insofar as all the more modern interpretations of the various quantum theories are ultimately based on the interpretation of SM, they depend for their validity on this interpretation.

Throughout the work the 'derivations' given are simply outlines of how more rigorous methods might be used.

Chapter 2

Basics: The Schrödinger Condition

"Did all of those old guys swing like that?"[1]

2.1 Method

Much of 20th century theoretical physics can be characterised by,

> Schrödinger (Maxwell, Dirac) Equation, how many ways can I write thee? Let me count the ways.

Mathematicians are notorious for ignoring the physical interpretation of their excursions into theoretical physics — sometimes brazenly so — and tend, like the Red Queen, to be capable of 'believing six impossible things before breakfast'.[2] The 'standard — Copenhagen — interpretation' of quantum theory is often taken for granted by mathematicians and philosophers, ignoring the fact that not all good physicists are good philosophers and it is common for philosophers to know very little physics. All too often no distinction is made between equations (which are an expression of a law of nature) and identities (definitions). There is, for example, more interest in the study of the Schrödinger (Maxwell, Dirac) *equation* than in the study of the physics of the real world. In fact,there is a long 'tradition' among mathematically inclined physicists dating from classical times that it should be possible to derive substantive scientific results from mathematics or logic alone.[3]

In this essay, I take the opposing view: that the physical interpretation of mathematics must be given greater importance than its structure, generality or even elegance (Boltzmann is reported to

5

have remarked that elegance should be left to tailors). This means, naturally, that some of what is written here may be controversial, indeed may, at times, seem a little pedestrian. But, I use as my excuse an essay by that great historian E. P. Thompson where he compares himself to that English bird the Great Bustard; stumbling about in the undergrowth of the real world trying to get airbourne in a pathetic contrast to some 'theorists' of history, who attempt to soar like eagles, way above the brute facts.[4]

There are several clear descriptions of the necessary conditions for a (physical) theory to be considered scientific; below is an extract from Lucio Russo's superbly written book[5]:

(1) *Their statements are not about concrete objects, but about specific theoretical entities.* For example, Euclidean geometry makes statements about angles or segments, and thermodynamics about the temperature or entropy of a system, but in nature there is no angle, segment, temperature or entropy.

(2) *The theory has a rigorously deductive structure.* It consists of a few fundamental statements (called axioms, postulates, or principles) about its own theoretical entities, and it gives a unified and universally accepted means for deducing from them an infinite number of consequences. In other words, the theory provides general methods for solving an unlimited number of problems. Such problems, posable within the scope of the theory, are in reality 'exercises', in the sense that there is general agreement among specialists on the methods of solving them and of checking the correctness of the solutions. The fundamental methods are proofs and calculation. The 'truth' of scientific statements is therefore guaranteed in this sense.

(3) *Applications to the real world are based on correspondence rules between the entities of the theory and concrete objects.* Unlike the internal assertions of the theory, the correspondence rules carry no absolute guarantee. The fundamental method for checking their validity, which is to say the applicability of the theory, is the experimental method. In any case, the range of validity of the correspondence rules is always limited.

Any theory with these three characteristics will be called a scientific theory.

Mario Bunge[6] has, perhaps, gone furthest with this approach.

What is missing from the list cannot be expressed so succinctly but is equally important. Perhaps most importantly, these criteria can easily be satisfied by purely phenomenological, even empirical theories; there is no doubt, for example, that these criteria are satisfied by Ptolomy's theory of the solar system.

Secondly, these criteria are notably, and I suppose intentionally, *static*; they give no indication of the *development* of the theory, how it relates to earlier theories. If a theory is to augment or replace an existing body of science or, in particular, if a theory is completely new, it must 'join up' in some sense with what might be called 'neighbouring' theories, that is, theories which are 'close' in the sense of dealing with related subject matters or theories about the same subject which have a simultaneous existence or have an historical concern with the subject.

- No theory of matter is of unlimited application; a theory of the structure and behaviour of some part of the natural material world must, in some way, make sense when viewed from the perspective of theories of the behaviour of 'adjoining' subject matter. The subject matter of theories of physics and chemistry and of chemistry and biology overlap and so chemists and physicists should be able to understand — and even use — each other's theories just as chemists and biologists are able to collaborate because their theories are mutually comprehensible. Theories of different parts of the physical world should not generate results (concepts or quantitative predictions) which are nonsensical or meaningless when viewed from a wider (or narrower) perspective.
- Equally important, no theory emerges from nowhere; no matter how novel or revolutionary a theory seems to be, the workers who generated that theory were always steeped in the existing theories which tried to describe the same phenomena and this is easily traceable in their work. Physical theories, even ones which seem

completely revolutionary, show lines of development from earlier work when it is possible to view them in a larger context.

It is this latter point, which is absent from the formal axiomatic approach but which is of supreme importance in thinking about the physical interpretation of new theoretical constructs. The danger is always that, if these connections are not made and a satisfactory physical interpretation is not discovered, theories may degenerate into semi-empirical recipes to be calibrated against known results and used to interpolate and extrapolate new facts; technologically handy, perhaps, but not science, no matter how mathematically sophisticated they might be. The real question is 'how is the science to be advanced and developed' not how sophisticated its mathematical articulation may be made to appear.

It is now common among theorists working in field and elementary particle physics to know very little about classical dynamical theories *and* not to be at all familiar with the work of the pioneers of quantum theory. This is extremely unfortunate from the point of view of the interpretation of the theoretical findings. Attempts to separate off quantum theory from its classical predecessor and its original roots by simply starting from a set of (usually uninterpreted) axioms generates a sense of arbitrariness in the physical understanding of the theories and a sterile *post hoc* approach to the justification of the axiom system; what I think of as the Ptolomaic method in science. Every new development in science — indeed in knowledge in general — always grows out of old concepts and a failure (or refusal) to understand this process impoverishes thought.

Partly for these reasons, I have not used some of the conventional shorthands of theoretical physics since I wish to emphasise the interpretation of the theory and the *size* of the quantities involved, not simply its structure. There will, therefore, be appearances of $\hbar$ and c, etc. in the text which are often set to unity to minimise clutter in the algebra. Also, I have tended to express myself in explicit a mathematical form: functions of space and time, since much is lost in over-abstract 'algebraic' formalisms.

Most of the symbols used are completely standard (ψ for state function, S for solutions of the Hamiltonian–Jacobi Equation (HJE), q^j for a coordinate, etc.), but I use some symbols which are either not standard or are not so familiar, here is a short list:

ρ is used as a generic symbol for a spatial (or space/time) distribution.

- With no subscript, ρ is the *probability of presence* distribution for a particle: $\rho = |\psi|^2$ which must integrate to unity in order to qualify as a probability distribution

$$\int_a^b \rho dV = \int_a^b |\psi|^2 dV = 1$$

where dV is the volume element and a, b are the limits of the relevant space.
- With a subscript, ρ_A (say) is used for the spatial distribution of some physical variable: for example ρ_E represents the distribution of energy E which is the value of E at a point in space multiplied by ρ. Clearly the mean value of this distribution will be

$$\langle A \rangle = \frac{\int_a^b \rho_A dV}{\int_a^b \rho dV} = \int_a^b \rho_A dV$$

H is the generic symbol for some kind of Hamiltonian:

- H^{CPM} is the Hamiltonian function of Classical Particle Mechanics.
- H^{SM} is the Hamiltonian function used in Schrödinger's derivation of the Schrödinger Equation (SE).
- $\hat{H}$ is the Hamiltonian operator of the SE.

Notice that the first two of these are *functions* and the last one is an *operator* as symbolised by its 'hat': $\hat{H}$. I have also used acronyms for many common phrases to avoid constantly repeating terms like 'Classical Particle Mechanics' and 'Schrödinger Equation' etc. throughout the text. Each symbol is defined where it first appears but, if the text is simply 'dipped into' at random, the symbols are better gathered together here.

2.2   Some Clarification

There are some extremely elementary points about non-relativistic quantum theory which have an important bearing on the interpretation of all quantum mechanical theories particularly when spin and special relativity are considered. It is, therefore, critical to have these points emphasised and clarified.

First of all it is essential to give a very brief outline of what the relationship of Schrödinger's Mechanics (SM) to CPM actually *is* since this relationship is at the heart of understanding the probabilities which occur in the SE and the meaning of many of the dynamical quantities involved. I use SM in preference to Quantum Mechanics (QM) partly because, like all those with a background in quantum chemistry, I have always needed the state functions for the computation of the all-important electron probability distributions which are as important as their energies in seeking explanations and mechanisms in chemistry. But my preference is much more broadly based than this. For reasons which will become clear as this chapter progresses I have not been able to find a coherent realistic interpretation of any of the other 'formulations' of QM. So, in a very real sense, I am using SM and its physical interpretation as my 'gold standard'; other formulations will be judged by how their interpretations stand up to that of SM. This will be taken up in later chapters.

Starting with Newton's original $F = \dot{p} = ma$, there are three formulations of CPM:

(1) Lagrange's Equations: This is the main technique for actually computing individual trajectories of particles in space particularly when non-Cartesian coordinates are used for that space.

(2) Hamilton's Canonical Equations: Here is Hamilton's method of using coordinates and momenta as the basic variables rather that coordinates and velocities used by Lagrange. Hamilton's method is again a technique for the calculation of particle trajectories in space.

(3) The Hamilton–Jacobi Equation: Derived by Hamilton and perfected by Jacobi, this equation provides a method of generating *all possible* trajectories for a given system in a given same space;

any individual trajectory can be obtained from the solution of this equation by providing some 'initial' conditions for the motion.

In summary, both Lagrange's and Hamilton's canonical methods generate a set of simultaneous *ordinary* differential equations for particle trajectories in a given space while the HJE involves a single *partial* differential equation whose solution is a function of all points in the relevant space. To emphasise this point, the coordinates (q^i, say) are the coordinates of the *particles* in the Lagrange and Canonical equations while the coordinates (again, unfortunately often using the same symbols q^i) are the coordinates of *points in space*.

Schrödinger started from the Hamilton–Jacobi Equation in spite of the almost universal opinion that Quantum Mechanics is based on Hamilton's canonical formulation.

2.3   Coordinates and Velocities

The idea that the classical limit of SM is the canonical equations of Hamilton rather than the HJE has some unfortunate effects. This point is worth looking at in detail here.

2.3.1   Coordinates

Hamilton's canonical equations are, in fact, sets of pairs of one *equation*

$$\frac{\partial H}{\partial q^j} = \dot{p}_j$$

which is the fundamental law of nature of CPM — equivalent to $F = ma$ — and one *definition*

$$\frac{\partial H}{\partial p_j} \overset{\text{def}}{=} \dot{q}^j$$

which *defines* velocity in the same way that

$$\frac{\partial L}{\partial \dot{q}^j} \overset{\text{def}}{=} p^j$$

defines momentum in the mechanics of Lagrange. The solutions of this system are $q^j(t)$ and $p_j(t)$ — the actual allowed *trajectories* and corresponding momenta of the particle(s). That is, in particular, the q^j are functions of *time* and they define a curve in space. So that, unlike the solutions of the SE and HJE, they are not the *independent* variables $(q^j; t)$. That is, to emphasise the obvious, Hamilton's canonical equations are not, despite superficial appearances, partial differential equations, but they are the *generators* of sets of simultaneous ordinary differential equations to be solved for the $q^j(t)$

An attempt at formal precision might look like:

- For the solutions of the canonical equation for a single particle

$$q^j : R^1 \to C \quad C \subset R^3$$

  where R^1 is the real number system (in this case modelling 'time') and C a subset of R^3 (modelling ordinary space E^3) which would normally be capable of being parametrised by $R^1 : C$ models a curve in ordinary space.
- While, for the q^j occurring in the HJE

$$q^j : E^3 \to R$$

  where E^3 is ordinary (Euclidean 3D) space and R is the real number system (assuming some kind of coordinate frame), so that abbreviating the first case to

$$q^j : R^1 \to R^3 \quad \text{(Lagrange, Hamilton)}$$

  and taking the liberty of (temporarily!) identifying E^3 and R^3

$$q^j : R^3 \to R^1 \quad \text{(Hamilton–Jacobi)}$$

  where, in this case, R^1 models a point in space.

This confusion, between the coordinates of a *particle* in space and those of a *point* in that space, is endemic in both non-relativistic and relativistic physics. It is perhaps at its most damaging in the above case where it reinforces the idea that the probabilities of SM refer to individual particles and not to points in space. This point has been emphasised over the years but to little effect; two quotes are sufficient to make the point:

An early one:

'It is important to realise the essential difference between the two modes of description. When we specify the system by charge and current densities ρ, $\boldsymbol{i}$ the space coordinates r_a and the time τ are independent variables, and the motions of the system are contained in the time variations in the functions ρ, $\boldsymbol{i}$. If we regard the system as composed of particles with charges e_k the space coordinates of these particles x_k are *functions of time*.'

Leon Rosenfeld, *The Theory of Electrons*, p. 28 (North Holland, 1951)

and, a more robust, contemporary one:

'In the physics literature a happy-go-lucky confusion between physical and geometrical coordinates is the rule.'

Mario Bunge, *Scientific Realism* (Prometheus Books, 2001).

A common example is the 'definition' of a point in spacetime as an 'event' even though there may be nothing at this point to take part in an event.

This unfortunate circumstance is made worse by the common practice of using the same notation for almost all coordinate systems; q^j is widely used for both the coordinates of a point in space which does not move and for those of the position of a particle in that space as $q^j(t)$ which does move. The best solution, in the case of the individual solutions of the HJE, for example when one has obtained a solution for a particular trajectory from $S(q^j, t)$ and some chosen initial conditions, is to use a different notation, $y^j(t)$ say, for the trajectory. This ensures that there is no confusion between the space and a trajectory in that space.[7]

2.3.2 Velocities

It is quite common to see Hamilton's definition of velocity

$$\frac{\partial H}{\partial p_j} \overset{\text{def}}{=} \dot{q}^j = \frac{dq^j}{dt}$$

used in a quantum mechanical context where the Hamiltonian *function* has been replaced by the 'Hamiltonian' operator of SM. But in both HJE and SE

$$\dot{q}^j = \frac{dq^j}{dt}$$

is as meaningless as dq^j/dq^k; points in space do not have a velocity any more than the coordinates which define them are interdependent.

This result obviously has some bearing on the interpretation of the Dirac and other, Schrödinger-like, equations.[8] Equations which are *linear* in momenta must have these momenta multiplied by some quantity which has the *dimensions* (LT^{-1}) of velocity in order that the whole expression can have the dimensions of energy. It remains to be seen whether or not these quantities are actually interpretable as velocities and, if they are, what are they the velocities of?

2.3.3   The Hamiltonian and the Energy

The Hamiltonian of CPM is a function of coordinates, momenta and, possibly, time, that is, it is a function of all space, all possible values of momenta and all possible times

$$H(q^j, p^j; t)$$

where the coordinate system q^j spans the space in which the dynamics occurs. In HJE, the momenta and energy for a particular point in space are given by

$$p^\ell = \frac{\partial S(q^j; t)}{\partial q^\ell} \quad \text{and} \quad E = -\frac{\partial S(q^j; t)}{\partial t} \tag{2.1}$$

where $S(q^j; t)$ is the solution of HJE. The condition that these momenta and this energy in the coordinate system $(q^j; t)$ actually satisfy the law of motion ('$F = \dot{p}$') is

$$H\left(q^j, \frac{\partial S}{\partial q^j}; t\right) = -\frac{\partial S}{\partial t} \tag{2.2}$$

which might be abbreviated to

$$H - E = 0 \tag{2.3}$$

that is, the pairs of coordinates and momenta (q^j, p^j) in the function H must be found by the *solution* of Equation (2.2) so that the Hamiltonian function is equal to the system's energy. Then and only then is it true that the Hamiltonian function is equal to the actual energy of the system and the q^j and p_j define the actual allowed motions — allowed by Newton's laws that is — of the system. To labour the point, only when the equations of motion are *solved* and the q^j and p_j become known functions of the time t (that is, they are coordinates of the *particle* not coordinates of points in space), does H become

$$H(q^j(t), p_j(t); t) = h(t) = E(t) \quad \text{(say)}$$

Notice that the form of the HJE shows that the momenta and energy are given by the action of partial differential operators acting on a function of the q^j and t of space and time was established by Hamilton and Jacobi long before Schrödinger's work.

2.4 Schrödinger's Method

Because Schrödinger's original paper is the widely cited source of SM and because its contents and method now seem to be forgotten, it is worthwhile to give a fairly full description of how Schrödinger approached the generation of what was originally called 'wave' mechanics. This is done in this section with the interpretation of the equations and their relationship to CPM very much in mind. The 'derivation' is done in terms of a general (orthogonal) coordinate system for reasons which will become clear later.

The high point of CPM is the equation of Hamilton and Jacobi; for a single particle in Cartesians it is

$$\left\{ \left(\frac{\partial S}{\partial x} \right)^2 + \left(\frac{\partial S}{\partial y} \right)^2 + \left(\frac{\partial S}{\partial z} \right)^2 \right\} + V + \left(\frac{\partial S}{\partial t} \right) = 0$$

where $S = S(q^k; t)$ is a function of 3D space and time (that is the q^k and t do *not* define the position of the particle) and whose derivatives with respect to these coordinates generate the momenta and energy of a particle at a particular point in space

$$\left(\frac{\partial S}{\partial x}\right) = p_x, \text{ etc. and } \left(\frac{\partial S}{\partial t}\right) = -E$$

When this equation is solved and the 'initial conditions' are available it can generate all the trajectories of the particle.

This equation was Schrödinger's starting point. His approach consisting of two surprising steps. First, he defined a new function ψ by

$$S = -ik \ln \psi \quad \text{i.e.} \quad \psi = \exp(iS/k)$$

where k must have the same dimensions as S — those of action (ML^2T^{-1}). He later found that, in order for his results to be in agreement with experiment, k must have the value of $\hbar$ so that

$$\psi = \exp(iS/\hbar)$$

This step is just a change in the *notation* in which HJE is expressed and seems to offer nothing new except to add complications to the solution of the HJE. The next step was, in accordance with the introduction of the imaginary unit i, to allow the function ψ to have the freedom to take on complex values and write the HJE in this new notation and then to give an entirely new method to solve the resulting equation for the complex ψ. It is this second step which actually effects the transition from CPM to SM, a completely new theory, in which both the real and imaginary parts of ψ can be given a physical interpretation, related to but not identical to that of the HJE.

The classical HJE for a single particle (the equation for N particles is simply obtained by changing the summation limit from N to $3N$) is

$$H^{\text{CPM}} - E = T + V - E$$

$$= \frac{1}{2m} \sum_{k,l=1}^{N} \left(g^{kl} \frac{\partial S}{\partial q^k} \frac{\partial S}{\partial q^l} \right) + V(q^j) + \frac{\partial S}{\partial t} = 0$$

where g^{kl} are the elements of the metric associated with the coordinate system q^j and the momentum components are expressed in the usual HJ way by

$$p_j = \frac{\partial S}{\partial q^j} \quad \text{and} \quad E = -\frac{\partial S}{\partial t}$$

and

$$S = S(q^j; t)$$

For emphasis, the 'Hamiltonian' in this equation has been written H^{CPM}. The potential energy[9] has been written as a function of spatial coordinates only but the derivation is still valid for those few cases where the equations can be solved when V has time dependence. In SM, we use the Schrödinger complex form of S which, when necessary, will be written

$$S'(q^j; t) = S(q^j; t) - iR(q^j; t)$$

where S and R are real. We then have:

$$T = \frac{1}{2m} \sum_{k,l=1}^{N} g^{kl} \left(\frac{\partial S'}{\partial q^k} \frac{\partial S'}{\partial q^l} \right) \tag{2.4}$$

$$= \frac{1}{2m} \sum_{k,l=1}^{N} g^{kl} \left(\frac{-i\hbar}{\psi^*} \frac{\partial \psi^*}{\partial q^k} \right) \left(\frac{-i\hbar}{\psi} \frac{\partial \psi}{\partial q^l} \right) \tag{2.5}$$

$$= \frac{\hbar^2}{2m(\psi^*\psi)} \sum_{k,l=1}^{N} g^{kl} \left(\frac{\partial \psi^*}{\partial q^k} \right) \left(\frac{\partial \psi}{\partial q^l} \right) \tag{2.6}$$

and

$$E = -\frac{i}{\psi} \frac{\partial \psi}{\partial t}$$

giving HJE in its Schrödinger form as

$$\frac{\hbar^2}{2m(\psi^*\psi)} \sum_{k,l=1}^{N} g^{kl} \left(\frac{\partial \psi^*}{\partial q^k} \right) \left(\frac{\partial \psi}{\partial q^l} \right) - \frac{i}{\psi} \frac{\partial \psi}{\partial t} = 0$$

Multiplying this equation by $\psi^*\psi$ gives, finally,

$$\frac{\hbar^2}{2m} \sum_{k,l=1}^{N} g^{kl} \left(\frac{\partial \psi^*}{\partial q^k}\right) \left(\frac{\partial \psi}{\partial q^l}\right) - i\psi^* \frac{\partial \psi}{\partial t} = 0$$

More compactly,

$$\left\{\frac{\hbar^2}{2m}|\nabla \psi|^2 + |\psi|^2 V\right\} - i\psi^* \frac{\partial \psi}{\partial t} = 0 \qquad (2.7)$$

Notice here that, if S had remained in its real (HJE) form the final multiplier $\psi^*\psi = |\psi|^2 = \exp(iS/\hbar)\exp(-iS/\hbar)$ would have been simply unity and the whole exercise pointless. However, with this exception, nothing new has been said, the solutions of this equation would contain nothing new. From the point of view of physical meaning, it is still just a change of notation.

Before proceeding to Schrödinger's solution of the above equation it is worthwhile to think what further information *could* possibly come from a complex S or ψ over and above the information provided by the solutions of the HJE for real S. After all, the HJE gives all possible trajectories when augmented by a set of 'initial' conditions for any given trajectory which is being sought: when an 'initial' coordinate is known and $\partial S/\partial q^j$ gives the corresponding p_j for some point in time.

The only thing we might reasonably expect, therefore, is some information about the possible 'initial' positions of the particle from the function R — the imaginary part of S' — to add to the momentum obtainable from the real S of S' simply because this is the only thing missing from a complete knowledge of the particle's motion. This does not seem possible but, and this is the key point, perhaps the use of S' can give some way of *dispensing* with that knowledge. In short and with considerable hindsight, it would not be surprising if the complex ψ gave information about the *distributions* positions of particles on trajectories, in some form or other since any actual initial position of a particle is an experimental specification not obtainable from theory.

If we seek a solution of a partial differential equation of the general form of equation of the general form of (2.7)

$$H(\psi) - E(\psi) = 0$$

we are looking for a solution ψ which satisfies the equation for all values of the coordinates of 3D space and time. That is (in Cartesians, for simplicity) for any values $(x_j, y_j, z_j; t_j)$, the expression

$$H(\psi(x_j, y_j, z_j; t_j)) - E(\psi(x_j, y_j, z_j; t_j)) = 0$$

must be true and, what is more any weighted sum of terms like the above must also be zero

$$\sum_j w_j(H(\psi(x_j, y_j, z_j; t_j))) - E(\psi(x_j, y_j, z_j; t_j)) = 0$$

(where the w_j are any numbers) and, in particular, the weighted sum of *all possible* values of $\psi(x_j, y_j, z_j; t_j)$ must satisfy this requirement, i.e.

$$\int (H(\psi(x_j, y_j, z_j; t_j))) - E(\psi(x_j, y_j, z_j; t_j)) dV dt = 0 \qquad (2.8)$$

should be true (where the integral is over all space and time and the weighting factors are the volume elements dV). This was possibly not the way that Schrödinger thought about the problem — we shall probably never know — he just presented his solution like the proverbial 'bolt from the blue'; this is offered as a feeble excuse for what follows!

But this, although not progress, since it is still simply playing with the solutions of the HJE, does suggest another approach to the problem of determining ψ. The above considerations do not represent any progress because they are nothing more than a *consequence* of HJE in new notation. Whatever obscure notation into which we change the form of the HJE, we are still solving an equation which contains Newton's law.

However, the fact that the HJE implies

$$\int (H(\psi(x_j, y_j, z_j; t_j))) - E(\psi(x_j, y_j, z_j; t_j)) dV dt = 0$$

the converse is false, i.e.

$$\int (H(\psi(x_j, y_j, z_j; t_j))) - E(\psi(x_j, y_j, z_j; t_j)) dV dt = 0$$

does not imply *only* the HJE.

There is a more general solution to the above problem which defines ψ; in addition to the trivial sum of many solutions to the HJE in its new notation given above, there must be solutions for which the expression

$$(H(\psi(x_j, y_j, z_j; t_j))) - E(\psi(x_j, y_j, z_j; t_j))$$

is *not zero*, that is there must be cases where, in the sum or integral, points in space where the expression is either positive or negative which, in the integral, *cancel*. That is, the class of solutions generated by the minimisation of $(H - E)$ which, at some points in space, do *not* satisfy the HJE in its ψ notation. This proves to be the solution that Schrödinger sought and which turns out to be the new law of particle dynamics in the microscopic region of masses and lengths.

> In other, more familiar, words, solving the variational problem of the minimisation of $(H - E)$ generates a new equation which is not the HJE from which Schrödinger started. This is a really bold step; what it means is that, in contrast to Newton's law as expressed in the HJE for the macroscopic region, there are regions in space where the HJE does not apply.

Fortunately, the variational problem suggested by Equation (2.8) is a classic problem in mathematical physics, first studied by Lagrange and Euler. Adopting the now-standard (since Lagrange) notation we write

$$\mathcal{L}(\psi, \psi^*, \partial_k, \partial_l, ; t) = H - E = \frac{\hbar^2}{2m}(\nabla\psi^*)(\nabla\psi) + \psi^*\psi V - i\psi^*\frac{\partial\psi}{\partial t}$$

where $|\nabla\psi|^2$ and $|\psi|^2$ have been written in full to emphasise that $\mathcal{L}$ is a functional of the two independent functions ψ and ψ^* and their derivatives. The condition for a solution ψ and ψ^* is

$$\int_a^b dV \int_{t_a}^{t_b} dt\mathcal{L}(\psi, \partial_k\psi, \psi^*, \partial_k\psi^*, \partial_l; t, \partial_t\psi) = \text{a minimum}$$

The solution to this variational problem consists of two parts:

(1) In the 'interior region' of the volume described by the limits of the spatial integration (symbolised simply by a and b above) a partial differential equation for ψ is given, generically known as the Euler–Lagrange equation

$$\frac{\partial\mathcal{L}}{\partial\psi} - \sum_{j=1}^{3}\left\{\frac{1}{\sqrt{g}}\frac{\partial}{\partial q^j}\left(\sqrt{g}\frac{\partial\mathcal{L}}{\partial(\frac{\partial\psi}{\partial q^j})}\right)\right\} - \frac{\partial}{\partial t}\left(\frac{\partial\mathcal{L}}{\partial(\frac{\partial\psi}{\partial t})}\right) = 0 \quad (2.9)$$

(2) Necessary boundary conditions are specified at the limits of the integrations which the function (ψ in this case) must meet at those boundaries.

Carrying out the variation for the function ψ^*, the expressions which must be supplied in this case are

$$\frac{\partial\mathcal{L}}{\partial\psi^*} = V\psi - i\hbar\frac{\partial\psi}{\partial t} \quad (2.10)$$

$$\frac{\partial\mathcal{L}}{\partial(\partial_j(\psi^*))} = \frac{\hbar^2}{2m}\sum_{l=1}^{3}g^{jl}\frac{\partial\psi}{\partial q^l} \quad (2.11)$$

$$\frac{\partial\mathcal{L}}{\partial(\partial_t\psi^*)} = 0 \quad (2.12)$$

When these expressions are substituted into Lagrange's general equation we have an equation which determines ψ in the 'interior' region

$$\left(V\psi - i\frac{\partial\psi}{\partial t}\right) - \frac{\hbar^2}{2m}\sum_{l=1}^{3}\left(\frac{1}{\sqrt{g}}\frac{\partial}{\partial q^j}\sqrt{g}\sum_{l=1}^{3}g^{jl}\frac{\partial\psi}{\partial q^l}\right) = 0$$

which can be arranged into a more recognisable form as

$$-\frac{\hbar^2}{2m}\sum_{l=1}^{3}\left(\frac{1}{\sqrt{g}}\frac{\partial}{\partial q^j}\sqrt{g}\sum_{l=1}^{3}g^{jl}\frac{\partial\psi}{\partial q^l}\right)+V\psi=i\hbar\frac{\partial\psi}{\partial t}$$

and shortened into the more familiar form

$$-\frac{\hbar^2}{2m}\nabla^2\psi(q^k;t)+V\psi(q^k;t)=i\hbar\frac{\partial\psi(q^k;t)}{\partial t}$$

which is SE for ψ in the general coordinates q^k and t which is usually written in 'operator form' as

$$\left(-\frac{\hbar^2}{2m}\nabla^2+V\right)\psi(q^k;t)=\left(i\hbar\frac{\partial}{\partial t}\right)\psi(q^k;t) \qquad (2.13)$$

$$\text{that is: } \hat{H}\psi=\hat{E}\psi \qquad (2.14)$$

Varying ψ yields an equation for ψ^* which is just the complex conjugate of the above.

There are a few preliminary points to be made about this equation and its derivation:

- Its source is the generalisation of the HJE and *not* from Hamilton's canonical equations.
- There has been no new quantum 'axiom' relating to the familiar empirical rule of making the transition from CPM to SM by using

$$p_{q^j}\rightarrow-i\hbar\frac{\partial}{\partial q^j}$$

since this rule occurs already in HJE. In fact, if Hamilton and Jacobi had been using the operator terminology, which is now *de rigeur*, they could have anticipated Quantum Theory by writing (in Cartesians)

$$\hat{p}_j\stackrel{\text{def}}{=}\frac{\partial}{\partial q^j} \quad\text{and}\quad \hat{E}\stackrel{\text{def}}{=}\frac{\partial}{\partial t}$$

The HJE would have been written

$$(\hat{p}_xS)^2+(\hat{p}_yS)^2+(\hat{p}_zS)^2+VS=-\hat{E}S$$

with, along with $\hat{x} = x$ etc. and therefore,

$$[\hat{p}_x\hat{x} - \hat{x}\hat{p}_x] = 1$$

looks very familiar today.

- However, what is new is that the momentum may now be complex since the derivatives of ψ include the original (HJ) terms from S plus new terms from the derivatives of R. This matter clearly needs further attention.
- The function ψ can be written as

$$\psi = \exp[i(S - iR)/\hbar] = \exp(iS/\hbar)\exp(R/\hbar) = f\exp(iS/\hbar)$$

where $f = \exp(R/\hbar)$ so that

$$\psi^*\psi = f^2$$

where both $f(q^j;t)$ and $f(q^j;t)^2$ need physical interpretation since R and hence f does not occur in the HJE, i.e. in CPM.

When the SE can be solved by standard methods the boundary conditions must be applied to the resulting solutions ψ. The boundary conditions are as important as the SE since they ensure the Hermiticity[10] of the operators $\hat{H}$ and $\hat{E}$, defined above, and therefore quantisation.

2.5 Interpretation

The solution $S(q^i;t)$ of the HJE is a function of the same *independent* variables $(q^i;t)$ as $\psi(q^i;t)$, the solution of the SE, that is, these variables are functions of *all* space and time. For both SE and HJE, the solutions refer to *all possible* particle motions which are compatible with the energy involved, so that the calculations resulting from the solutions of the SE do not refer to the motions in a single particle trajectory. That is to say, *all* points in space are on some trajectory which satisfies the HJE for some 'initial' conditions and the spatial distribution of these allowed trajectories are fixed by the functions S or ψ.

The relationship of the HJE to the SE is made even more convincing by Poincaré's interpretation of Jacobi's last multiplier of Hamilton's equations — a constant — as indicating that the probabilities of all the possible solutions for a given energy are equal.[11] If the probabilities of occurrence of all possible solutions of Hamilton's canonical equations are all equal then, obviously, the HJE which generates all these possible solutions will generate them all with equal probabilities.

> This carries over into the interpretation of the SE whose solutions therefore refer to the set of all possible particle trajectories all of which have equal probability of occurring in the probability distribution obtained by the square of the magnitude of the state function for the particular case in hand.

Having said this, it is difficult to see how any other case would arise; it seems impossible to imagine a case when some particular trajectory would be chosen over any other in the HJE or the SE when all occur for the particular set of circumstances — same dynamical law, force field and boundary conditions. However, from the point of view of the connection between CPM and SM, this work of Poincaré is the first appearance of probability in the interpretation of Hamilton and Jacobi's work in bringing CPM to its final form. Just as the interpretation of the gradient components as momenta predated SM, so did the probability interpretation of the final form of CPM.

This result, that the *referent* — what SM actually describes — is the set of all possible trajectories for given circumstances, is absolutely crucial for any interpretation of SM. If an attempt is made to interpret SM as referring to the *individual* particle trajectories (i.e. coordinates of presence for a particle), then it is quite literally true that 'one does not know what one is talking about' since the state function ψ is a function of space and time and does not contain any dependence on particle coordinates.

2.5.1   The Physical Interpretation of $|\psi|^2$

I take the view that Kolmogorov is right and that probabilities are indeed relative measures of sets and that statistical measurements (or verifications) of probabilities are nothing more or less than approximations to these measures obtained by experimental means. This is to say that:

Probabilities are *real* numbers obtained theoretically from the measures of subsets of a given set. These real numbers are not associated, *per se* with randomness in any way; indeed, in simple cases they may be uniquely determined by simple analytic methods or even by counting.

Statistics is a collection of mathematical techniques developed to assist in getting the best results from a series of experimental measurements and generates a pair of *rational* numbers: a best value and an error bound.

There is no *direct* relationship between probabilities and statistics; statistics can be used to obtain the best ('most probable' is the unfortunate colloquial usage) value of a single quantity, for example the speed of light, where no probabilities are involved; the speed of light is a fixed quantity and statistics can help in the analysis of experimental determinations of such quantities. The randomness involved in making statistical measurements of (relative) probabilities — relative measures of subsets — occurs because one must try one's best to obtain results which are not prejudiced, and random measurements are the simplest way to ensure this.

Similarly, probabilities do not have to be associated with non-deterministic phenomena; it is perfectly possible, and useful in some cases, to calculate probability distributions for completely deterministic systems like a simple pendulum.[12] In the early days of QM, there was much concern that the occurrence of probabilities in the theory implied that the underlying *motions* were random[13] and there was a failure of causality; this is nonsense, it is the *measurements* which must be random to ensure unprejudiced reliability! Appendix B has a discussion of this question.

The Kolmogorov axioms apply to measures of subsets of a given set. That is, for a set Ω and subsets W_j a *measure* function

$$P : \Omega \to R, \quad P : W_j \to R$$

is defined from the subsets $W_j \subset \Omega$ to the real numbers such that

- the probability of the larger of two sets is not less than that of the smaller

$$P(W_1) \geq P(W_2) \quad \text{if } W_1 \subset W_2 \tag{2.15}$$

- the sum of the probabilities of two *disjoint* sets is the sum of their individual probabilities

$$P(W_1) + P(W_2) = P(W_1 + W_2) \quad \text{if } W_1 \cap W_2 = 0 \tag{2.16}$$

This result may be extended by recursion to any *denumerable* number of subsets of Ω.

- the probability of the enclosing set is unity

$$P(\Omega) = 1 \tag{2.17}$$

It is, perhaps, worth noting here that there is both a conceptual and a considerable practical difference between the probability of *occurrence* of a phenomenon and the 'probability' of *measuring* that occurrence. The former is a property of that part of the physical world[14] being studied while the latter, as well as depending on the physical world, depends on the instruments being used, the theory of operation of those instruments, the skill of the experimenters and a host of other, practical, unquantifiable, matters.

In the case of probabilities generated from a probability density which is a function ρ from the set X (members $x \in X$, subsets $X_j \subset X$) onto the real numbers, the corresponding results are

(1)

$$\int_{X_1} \rho(x)dx \geq \int_{X_2} \rho(x)dx \quad \text{if } X_1 \subset X_2$$

(2)

$$\int_{X_1} \rho(x)dx + \int_{X_2} \rho(x)dx = \int_{X_1+X_2} \rho(x)dx \quad \text{if } X_1 \cap X_2 = 0$$

(3)

$$\int_X \rho(x)dx = 1$$

which may always be arranged by multiplication by a numerical factor $1/N$ if

$$\int_X \rho(x)dx = N < \infty$$

Now, to qualify mathematically to be a distribution function a function ρ must be:

(1) Single valued
(2) Non-negative
(3) Integrable to a finite value

Thus, if

$$\rho(x) = |\psi(x)|^2$$

the square of the norm of any single-valued function has the correct *mathematical* requirements to be a probability distribution function. This is precisely the interpretation of the state function of SM. These measures are probabilities of *presence* for a particle or relative probabilities of particles for many-particle state functions.

Especially important in the connection between the SE and the HJE, the only way to obtain the SE which is valid for any system of coordinates, is the method which Schrödinger actually used: a variational technique based on the substitution, in the HJE, of

$$\psi = \exp(-iS/\hbar) \tag{2.18}$$

which leads to the expression

$$p_j = \frac{\partial S}{\partial q^j} = -i\frac{\hbar}{\psi}\frac{\partial}{\partial q^j} \tag{2.19}$$

This expression gives, in the same way as the classical expression gives, only the *value* of the particle momentum at a point in space (and time); in order to have the distribution (or *density*) of momentum in space and time this value must be multiplied by the probability distribution of the particle in space and time

$$\text{Momentum density} = \psi^*\psi \times \left(-i\frac{\hbar}{\psi}\frac{\partial\psi}{\partial q^j}\right) = -i\hbar\psi^*\frac{\partial\psi}{\partial q^j} \qquad (2.20)$$

In spite of these considerations about the way to derive the SE, there is an almost universal opinion that the quantum Hamiltonian *operator* and hence the SE can be derived from the classical Hamiltonian function by using a simple substitution

$$p_j \to -i\hbar\frac{\partial}{\partial q^j}$$

There are two mistakes in this familiar, innocent-looking recipe apart, that is, from the obvious fact that it does not work[15]:

(1) The operator $\hat{H}$, occurring in the SE and related equations is simply *called* the Hamiltonian operator; in reality it is just one of the two Euler–Lagrange equations resulting from the variational derivation of the SE. The substitution recipe does work in one very particular case — if the q^j are Cartesian coordinates — because, only in Cartesians is[16]

$$\nabla^2 = (\nabla)^2$$

In all other systems of (in practice orthogonal) coordinates, the correct Hamiltonian operator can only be *derived* by the variational procedure. In fact, the correct place to make the substitution given by Equation (2.5) is in forming the quantum-mechanical *function*, from the Hamiltonian *function*, that is the kinetic energy is

$$T = \left|\frac{\hbar}{2m}\nabla\psi\right|^2 \qquad (2.21)$$

(2) However, since the functional involved in the variational technique is invariant against transformations of spatial coordinate systems, once the operator ∇^2 has been obtained for one coordinate system it may be easily obtained for any other. That is, the correct operator may always be obtained using the simple recipe

of transforming ∇^2 from Cartesians to the coordinate system in use. This latter method is just a convenient rule of thumb; it is not part of a coherent, interpreted theory, and the operator $(-\hbar^2/2m)\nabla^2$ does *not* have the physical interpretation as kinetic energy and, actually, still awaits a physical interpretation. When the class of functions ψ_j ensures that this operator is Hermitian the *mean values* of $\psi_j^*(-1/2)\nabla^2\psi_j$ and $(1/2)|\nabla\psi_j|^2$ are the same but the *distributions* differ by $(1/4)\nabla^2\psi_j^2$. These matters will be taken up in more detail in Section 4.1.

Most importantly, using this prescription encourages the view that SM is just an arbitrary formalism — albeit generating extremely useful energetics and probability distributions — and does not have its own dynamical law which, in its domain of applicability, replaces Newton's law. In fact, Schrödinger's variational method shows that SM has a very clear relationship to CPM. But the substitution recipe via Cartesians can always be used a shortcut to avoid the rather tedious variational derivation for every curvilinear coordinate system. This sordid shortcut will be used in some of what follows.

Some of the confusions in the relationship between CPM and SM may be avoided by using an extended nomenclature (for a single particle for simplicity):

- The CPM Hamiltonian function is given by

$$\frac{1}{2m}\sum_{j=1}^{3} p_j^2 + V$$

  which, in the HJE context, becomes

$$\frac{1}{2m}(\nabla S)^2 + V$$

- The SM Hamiltonian function is given by

$$\frac{\hbar^2}{2m}|\nabla\psi|^2 + V\psi^*\psi$$

  where

$$S' = -i\hbar\ln(\psi)$$

- The SM Hamiltonian operator, formed in the Euler–Lagrange equation by Schrödinger's method is given by

$$-\frac{\hbar^2}{2m}\nabla^2 + V$$

The two *functions* have a direct physical interpretation while the operator does not.

2.5.2 Classical and Quantum Conditions

Although the HJE *refers* to the totality of particle trajectories given by a particular example of CPM, in order to obtain a specific one of the infinite number of such trajectories or a family of them we must

- find a coordinate system in which the 3D partial differential equation will separate into three ordinary differential equations.
- specify some 'initial' conditions for the trajectory (q_0^j, p_{i0} say).

This is not possible in the sub-atomic case; certainly in practice and, according to current opinion, even in principle. So one has to fall back on the assumption — strengthened now by Poincaré's interpretation of Jacobi's last multiplier (see Appendix A) that each trajectory is as likely to occur as any other. That is, all possible initial conditions may be assumed which are sufficient to satisfy the condition that all possible trajectories are present *with equal weight*. This is the implicit assumption in Schrödinger's extension of Jacobi's theory. One might say, rather unkindly, that Schrödinger made a virtue of necessity in being able to ignore the problem of initial conditions. This condition is also central to Statistical Mechanics and provides a theoretical grounding for the arbitrary assumption of 'equal *a priori* probabilities'.[17]

> The obvious result of these considerations is that both the SE and the HJE *refer* to the same thing: the set of all possible trajectories consistent with the conditions of the dynamical law; the HJE in the classical case and the SE in the quantum case.

2.6 Measurement

As it is presently constituted SM, like all other physical theories, says nothing about experimental measurements. The variables occurring in the theory are those relating to the particles or fields being described; there are no coordinates or operators for the measuring apparatus, the observer or his mind. For example, the solutions of the SE for a hydrogen-like atom or ion simply give allowed energies and electron distributions for the states of the system. However hard one looks, the design for a UV spectrometer is nowhere to be found among these solutions. The historical obsession with a 'theory' of measurement is a legacy of positivism and instrumentalism; two outdated approaches to the philosophy of science long ago shown to be inadequate and which have been abandoned by all philosophers and only live on as ghosts in texts on QM.

In this context, it is worth saying a few words about the mysterious 'Uncertainty Principle', often written as

$$\Delta q \Delta p \geq \frac{\hbar}{2}$$

where Δq and Δp are often not defined unambiguously, sometimes accuracy of *measurements* of q, p and sometimes the (unmeasured) *values* of coordinate and momentum. This principle illustrates many of the points of confusion about both the existence and measurement of the simultaneous position and momentum of (say) a particle. Firstly, this relation is not autonomous, it is a theorem in SM, not a principle and to emphasise this it is best written as

$$\sigma_q \sigma_p \geq \frac{\hbar}{2} \tag{2.22}$$

where, for a given state function and its associated probability distribution, σ_q and σ_p are the standard deviations of the coordinate and momentum distributions. Secondly, it says nothing about the limits to the measurability (simultaneous or otherwise) of coordinates and momenta, it simply places a limit of the product of the distributions of these quantities. In fact, the most uncertain thing about this principle is how to interpret it. Since the referent of SM is the set of all possible trajectories for given circumstances, the principle cannot apply to a *single* particle; the pair (p_j, q^j) in the theory refer to a point in space not those of a particle trajectory. Even if it were possible to measure these two quantities simultaneously this would simply be a single measurement from which nothing about the probability distributions involved could be deduced.

In the general case of the relationship of measurement to the theoretical value for concreteness, consider the measurement of some real-valued variable of a system X; let T be the theoretically computed quantity then the theoretical result is

$$T : X \times U \to R$$

where U is a system of units and R is the real number system. However, the corresponding experimental magnitude (T') (the value of T for a specific concrete object) might be expressed as something like

$$T' : X \times U \times M \times M' \times E \times O \times \cdots \to P(R)$$

where M is the set of methods of measuring T, M' is the set of theories about how these methods work, E is the set of equipment used, O the set of experimental workers, etc. and $P(R)$ is the set of intervals in the real numbers with rational end points, the best value and an error bound. Clearly T and T' are different mappings with different domains and different ranges and the relationship between them, however intuitively obvious, is not at all mathematically simple; it would be a gigantic exercise to have a theory for one method of measurement and completely impossible to produce a general theory of measurement.

2.7   Schrödinger's Mechanics, Axioms and the Schrödinger Condition

In Section 2.2, some of the deficiencies of the formal axiomatic method were mentioned in the context of understanding the interpretation of physical theories. It has already been mentioned that, in SM, some of the axioms needed in other formulations of QM are theorems.

SM takes over all the concepts and interpretations of CPM *except* the fundamental dynamical law.

The CPM law $(F = \dot{p})$ in its most general form is HJE which may be simply expressed as

$$H^{\mathrm{CPM}} - E = 0$$

The allowed trajectories of the particles are the ones which make the *classical* Hamiltonian function equal to the energy.

While the corresponding law in SM is

$$\int dV dt \left(H^{\mathrm{SM}} - E \right) = \text{ minimum} \qquad (2.23)$$

where the notations H^{CPM} and H^{SM} have been used to distinguish the classical and quantum Hamiltonian *functions* defined on page 29.

> The minimum condition in Equation (2.23) has already been described as the Schrödinger Condition (SC) since it is the overarching condition from which equations which govern the energetics and particle distributions for any (micro-scale) system.

The allowed distributions of particles are the ones for which the mean value of the spatial distribution of the *quantum* Hamiltonian function are equal to the mean value of the energy distribution.

With this law and the associated interpretation, the following are easily *derivable*:

(1) The commutation rules for q^j and p_j are trivial to demonstrate, in fact, they go back to HJE.
(2) The Heisenberg 'uncertainty' principle is not a principle but a theorem relating the probability distributions of a conjugate pair of coordinates and momenta.
(3) The rule for forming the mean value of a dynamical variable.

These facts flow naturally from SM. But, most importantly, *no* other formulation of QM gives the all-important state functions — they are either left as formal objects or smuggled in from SM.

2.8   Limiting Cases and Trajectories

Both of the major innovations of the 20th century in physics — Special Relativity (SR) and QM — are said to go over into Newtonian mechanics under suitable limiting conditions; v/c small and very small $\hbar$, respectively. It is useful to compare these two limiting cases.

In the case of SR it is very easy to show that this is the case because SR is not a new *dynamical* law; it is just a restriction on the applicability of Newton's second law

$$F = \frac{dp}{dt}$$

which, in Cartesians, becomes

$$F = \frac{dp}{dt} = \frac{dmv}{dt} = m\frac{dv}{dt} = ma$$

since Newton made the 'self-evident' assumption that the mass of a given body was a constant and, in particular, did not change with time. If, as Einstein insisted, the measured mass of a body depended on its velocity, Newton's law still holds but its reduction to its familiar form, $F = \dot{p}$ is no longer valid:

$$\frac{dp}{dt} = \frac{dmv}{dt} = m\frac{dv}{dt} + v\frac{dm}{dt}$$

and when the relativistic rules are imposed, generates a modified expression which for small v/c does indeed regenerate $F = \dot{p}$.

The key point here is that SR *constrains* Newton's law, it does not *replace* it. This means of course, that the *referent* of theory is still the same; the solutions of either equation is the *trajectory* of a body.

The situation in SE is quite different. Firstly, because while in SR v is a dynamical quantity, a property of the body, $\hbar$ is just a number. The *dynamical law* of SE would not be changed by changing the value of $\hbar$, all that would happen is the values of the predicted quantities would be in error. Secondly and much more importantly, the dynamical law of SE is different from Newton's law and therefore the referent of the two theories do not necessarily coincide.

Recall that we are working entirely with interpreted SM which takes over the physical interpretation of the HJE together with the interpretation of the state function as the spatial *probability* distribution of all possible trajectories associated with a given energy. That is, *whatever the numerical value given to the symbol $\hbar$*, the referent of the theory is not the trajectory of a body but a probability distribution of all possible trajectories for the given conditions and in all the space covered by the coordinate system. This means that SM and CPM or both autonomous sciences, neither can be reduced to the other. For example SM is powerless to describe the motion of the moon around the earth; all it can do is calculate, for a given energy, some properties of the set of all possible orbits of the moon around the earth. Thus, in no sense is SM more fundamental[18] than CPM; they are separate sciences dealing with different aspects of the world.

One can find in the literature expansions of the SE in terms of powers of $\hbar$ which, when terms in $\hbar^2$ and higher are neglected, give an equation similar to the HJE. At first this looks rather surprising since Schrödinger took the HJE as his starting point, 'simply' changing the notation by assuming that

$$S = -i\hbar \ln(\psi)$$

In other words, one might think that the HJE should be immediately recoverable with this change of notation. But there is evidently some 'hysteresis' in this apparently reversible transformation. Schrodinger changed the Hamiltonian *function* from its classical form into the QM Hamiltonian *function* and then used this form in his variational derivation of the SE. But the limiting procedure above starts not from the quantum Hamiltonian *function* but from the Hamiltonian *operator* which is the object which involved in the SE — the Euler–Lagrange equation of the variational problem — hence the necessity of the higher-order terms.

Finally a word about particle trajectories from a different angle: from the point of view of probability distributions. SM provides no information at all about individual particle trajectories since, unlike the HJE of CPM, there is no possibility of obtaining anything corresponding to the initial conditions of such trajectories. SM is a probabilistic theory, its referent is the set of all possible trajectories for given conditions. In any non-trivial case this is perfectly obvious; whatever the numerical value given to $\hbar$ the solutions of the SE for the H atom are always 3D functions of space, not curves in that space.

> This does *not* mean that the particles do not have trajectories; all that it does mean is that SM cannot tell us what they are.

This latter point is a common feature of *all* probabilistic theories; there is no way of obtaining information about the specific items in any probability distribution:

(1) It is common knowledge that one cannot use probability theory to predict the result of dice tossing. Not the least reason for this obvious result is the fact that, when the probabilities were designed and 'computed' there is no mention made of the way in which these probabilities were to be subject to experimental verification. The equality of the six possible probabilities is basically due to the symmetry of a cube and does not mention the properties of any concrete example of a cube. There is no mention, for example, that any real cube used in tossing experiments should be of homogeneous mass density (not 'loaded') and should be tossed from a height that is large compared to its dimensions.

(2) When one computes the probability distribution for the position of a pendulum on its path there is no way that the actual time-dependent trajectory can be found from this distribution because one would need additional information which is not available from the distribution in addition to some initial conditions for the motion.

(3) The referent of the QM probability distributions is the set of all possible trajectories in the available space. Even if one could single out a particular path in space — which is not possible, whatever the size of $\hbar$ or the magnitude of any quantum numbers — one would still need initial conditions to specify the trajectory along this path.

(4) The 'converse' of the statement in item 1 is also true; it is equally true that the result of a single measurement cannot give any information about a probability distribution. One can draw no conclusions about *any* probabilistic theory from either non-random measurements or from a single measurement.

The other side of these considerations is that, since probability distributions do not refer to individual events, any experimental setup to verify or test QM probabilities must consist of many *random* measurements on many individual examples of the objects of the distributions. Successive measurements on the same system are not random and not a valid way of putting QM to the test. In spite of this, one particular formulation of QM (not SM) claims to be able to generate classical trajectories from quantum theory. We shall need to examine this claim in some detail later in Section 3.5.

One final point on measurement: this work is concerned with SM and this entails the probability interpretation of $|\psi|^2$ as well as the numerical values of physical properties computed from ψ. SM therefore makes no claim to be able to predict the results of individual measurements of physical properties other than those for which the distributions are homogeneous: the eigenvalues of the SE and operators which commute with the Hamiltonian operator.

Endnotes

1 The young John Coltrane in the 1960's, on first hearing a recording of Sidney Bechet made in 1932.

2 From the 'wave-particle duality' via 'complementarity', onwards to the 'collapse of the state function or wave packet' through 'Schrödinger's cat' and 'virtual particles' to 'Multiple Universes'.

3 This point, together with some other 'philosophical' matters are taken up in Appendix C.

4 *The Poverty of Theory* Merlin Press, 1978; he was, of course, aware of the near-pun in his comparison. And, like the famous historian who says that both he and the Great Bustard can only get airbourne in the face of strong headwinds, I anticipate plenty of opposing forces.

5 *The Forgotten Revolution* English edition, Springer, 2003.

6 See, for example, his *Foundations of Physics* Springer-Verlag, 2004.

7 This is used in practice in Chapter 5.

8 That is, equations like the SE , not equations like Schrödinger.

9 Notice, in line with the emphasis on the fact that both S and ψ are functions of *space* — not of a particle's position in space — it should be emphasised that the potential *energy* is the product of the potential *function* of space and some property (charge, say) of the particle.

10 It is worth noting that there is no such thing as an operator which is *per se* Hermitian, only when the space of functions on which it operates is specified can it be shown to be Hermitian. So, when one sees, in words reminiscent of the first chapter of the book of Genesis, 'Let $\hat{A}$ be a Hermitian operator' always be on guard.

11 See Appendix A.

12 It is, for example, of little practical use to have a precise knowledge of the time evolution of a system if measurements can only be at random times when the phenomenon 'turns up'.

13 This may have been the reason for Schrödinger's original opinion that his theory was not about probabilities.

14 More precisely the current *model* of that part of the physical world.

[15] Try it for spherical polars or prolate spheroidals, say.

[16] There is obviously a strong case for the routine use of another notation for ∇^2 to avoid this type of confusion; Δ is used by some authors.

[17] Indeed, it is not much of a typographical change to replace '*a priori* probabilities' by '*a Poincaré* probabilities' and it has the satisfaction of removing an unnecessary axiom from the theory.

[18] Some extreme reductionists might challenge this claim but reductionism is not the same as 'constructionism'. That is, although it may be possible to claim that the 'high-level' sciences must *obey* the laws of the 'lower-level' (said to be 'more fundamental') sciences involved, the behaviour of the higher-level subject matter cannot be derived from the low-level laws. A good example is the tail display of a peacock; when this happens all the processes involved in the action must obey the laws of physics and chemistry but a knowledge of physics and chemistry would be useless in an attempt to predict this activity. One might describe the low-level laws as 'constraints' on what the biology is trying to do. Different laws emerge for each level of phenomena.

Appendix A: Jacobi's Last Multiplier

The material in this appendix is not submitted as any kind of proof that the referent of the SE is the same as that of the HJE, just as an indication of how close the past masters of CPM came to developing the foresight and tools essential for the understanding of SM. Things would have been much simpler if, to refer to the motto of this chapter it had been realised how well 'the old cats could swing'.

A.1   The Multiplier

In his study and extension of Hamilton's work, Jacobi defined a new quantity associated with systems of ordinary linear differential equations like Lagrange's and Hamilton's canonical systems: it has become known as Jacobi's Last Multiplier and is usually given the symbol M.

In his original paper, he studied a system of such equations more general than those of Hamilton in that he allowed for forces which were not derivable from a potential, in particular forces which may be velocity dependent. In this short *resumé*, reference will be made to only Hamilton's canonical equations:

$$\dot{q}^j = \frac{dq^j}{dt} = \frac{\partial H}{\partial p_j} \quad \dot{p}_j = \frac{dp_j}{dt} = -\frac{\partial H}{\partial q^j} \tag{A.1}$$

In the notation of the time, Jacobi 'separated the differentials' of the time derivative and wrote

$$dt = \frac{dq^j}{\frac{\partial H}{\partial p_j}} \quad dt = \frac{dp_j}{-\frac{\partial H}{\partial q^i}} \tag{A.2}$$

So that the whole $2n$ of the ordinary differential equations may be summarised as

$$dt = \frac{dq^1}{\frac{\partial H}{\partial p_1}} = \frac{dq^2}{\frac{\partial H}{\partial p_2}} \cdots = \frac{dq^n}{\frac{\partial H}{\partial p_n}} = \frac{dp_1}{-\frac{\partial H}{\partial q^1}} = \frac{dp_2}{-\frac{\partial H}{\partial q^2}} \cdots = \frac{dp_n}{-\frac{\partial H}{\partial q^n}}$$

Jacobi generalised Euler's 'integrating factor' (or multiplier) for the solution of ordinary differential equations to the general case of sets of simultaneous differential equations. The required factor, M, is then a solution of

$$-\frac{d\log M}{dt} = \sum_{j=1}^{n} \left(\frac{\partial^2 H}{\partial p_j \partial q^j} - \frac{\partial^2 H}{\partial q^j \partial p_j} \right)$$

But Hamilton's Equations means that each of the terms in this sum is zero so the total sum gives

$$\frac{d\log M}{dt} = 0 \quad \text{that is } M = \text{constant}$$

Jacobi's last multiplier (M) actually satisfies an equation of surprisingly familiar form; if we take Cartesian coordinates (x, y, z) and three functions of these Cartesians $(f, g, h$, say, which satisfy a set of three differential equations of the type to which Jacobi's theorem applies) then the multiplier (in general $M(x, y, z)$) satisfies

$$\frac{\partial Mf}{\partial x} + \frac{\partial Mg}{\partial y} + \frac{\partial Mh}{\partial z} = 0$$

which is recognisable as, for example, as the equation determining the continuity equation of a fluid in which (f, g, h) are the velocity components at (x, y, z) and M is the density of the fluid at that point. Now, in general, one should proceed with extreme caution when comparing the properties of different physical systems whose behaviour involves similar equations, but in this case it does prove extremely suggestive.

The referents of this form of equation are known to be

(1) the conservation of a fluid of spatial density $\rho(\boldsymbol{r})$, in time,
(2) the conservation of something involving the particle trajectories which obey Hamiltonian mechanics,
(3) the conservation of the spatial probability density $|\psi(\boldsymbol{r})|^2$ of particle trajectories obeying SM.

In the 19th century, of course, nothing was known about the last of these equations but Boltzmann, Larmor and Poincaré noted and commented on the first two. In particular, Poincaré was struck by the fact that, for particles in the *phase space* of Hamiltonian dynamics the multiplier M (potentially a function of space and time) is, in fact, a *constant*. He used the analogy with the flow conservation equation to say that the constancy of M meant that *in the phase space* (q^i, p_i) all the trajectories were equally likely.

In order to appreciate this interpretation one must remember the derivation of the equation which determines M; it is a linear combination of *all* $6n$ Hamiltonian equations for n particles, that is, and this is the key observation, M applies to *all possible* trajectories in *phase* space.

Poincaré was very disturbed by the implications of his result when combined with the very early indications that energy and other dynamical quantities might be *quantised*. Obviously, in this case, whatever the equivalent of M was to be it could not be a constant; much worse, since quantisation specifically excluded certain motions, the 'quantum M' could not even be continuous; it must consist of several separate parts with regions of zero in between. Now, for someone working in Poincaré's lifetime this must have seemed a disaster since it meant that the principle tool for solving theoretical problems in physics — differential equations — could no longer be used.

What Poincaré could not have forseen was that, albeit unwittingly, Jacobi had already provided a way out of this apparent *impasse* in the form of his greatest contribution to classical mechanics: his simplification of Hamilton's results to give the HJE. This equation has two huge advantages over the Hamilton canonical equations in this area of interest: it is an equation involving only the $3n$ *spatial* coordinates and this one equation can generate *all* the trajectories of the n particles in ordinary space. And, one might say in a final touch that Schrödinger's use of the HJE translates this equality of probabilities of trajectories into the distribution of the particle having these trajectories in *configuration* space.

Only 14 years after Poincaré's death in 1912, in Schrödinger's hands, the HJE was used to develop a new mechanics which had the following features:

- It preserved the central role of differential equations in physics by giving an equation which generated quantisation.
- When solved, this differential equation gave a solution whose referent, for given values of the quantised dynamical variable (often energy), is all possible spatial motions of the particles.

More than this, the spatial distribution of the set of all trajectories is a measurable set (normalisable) and its subsets satisfy all the requirements of Kolmogorov's formal probability theory as has been outlined in this chapter.

Unfortunately, this historical and logical development from Jacobi via Poincaré to Schrödinger seems, in the 20th century, to have been forgotten or overlooked[a] and has become lost amongst a plethora of colloquial interpretations of probability which have become part of the 'standard interpretation'.

---

[a]It seems incredible to me that Schrödinger's original ('Lagrangian') method of obtaining his all-powerful equation only appears in one text currently in print, and this is a reprint of a book originally published in 1929.

Appendix B: Probability and Statistics: An Example

If the argument in Section 2.5 does not sound satisfactory it is worth a detour into a very simple example: SHM in 1D. This problem is completely soluble using any of the formulations of CPM; the trajectory is completely determined by the solution of the relevant equation and some 'initial' conditions. It is also straightforward to calculate the probability of presence of the oscillation particle on its trajectory. As expected, the probability of presence function does not depend on any initial conditions. There are two cases to consider if we wish to perform measurements to verify these two results:

(1) In the first case we can simply calculate where the particle will be at any chosen set of times and look for it at those times. If the experimental results coincide with the CPM predictions then that is all that need be done.

(2) The second case requires a completely different approach. One method is to choose a set of points on the trajectory (known of course, because the motion is in 1D) and count the numbers of times the particle is found there for each case. Then, if we have taken enough points and the same number of 'measurements' at each point, then the ratios of these numbers should match the ratios of the predicted probability of presence at those points.

However, consider the following situation where we are much less in control of the situation and able to set up apparatus easily to verify the theories. Suppose that we know or suspect that the particle is executing SHM but we cannot get close to it or even see it and our only clue to the motion is that, at completely random times, it emits a flash of light from one or other of the points on its trajectory and we can determine the point from which the each flash came.

This time there is only one usable theory; the exact calculation of the trajectory cannot be done because *we do not know the initial conditions*. Even if we did know these conditions, we cannot compute the trajectory from our measurements since they occur at *random* times.

These statistical data are presumed to be random enough to make the counts of the flash positions a good enough approximate 'measure' of the relative cases to be able to predict the calculated probability-of-presence measures. What is important here is that it is the statistical *measurements* must be random to ensure a satisfactory and unbiased set of relative measures; the SHM *motion* is perfectly deterministic but *not known to us*.

This case, although artificial and based on CPM, has strong similarities to the quantum case in the sense that we have no knowledge of initial conditions for any of the (usually billions of) particles in the system and we have to be satisfied with such data that we can glean 'as it turns up'. This necessity of randomness in the experimental observation contains absolutely no information about whether or not there is randomness in the underlying motions in CPM or SM. However, it is hard to see how one could have any satisfactory measurements on a system undergoing random changes using techniques which depend on random measurements completely independently of any randomness in the system's motions.

With these considerations in mind, we should note that the theory which goes under the name 'Statistical Mechanics' should really be called 'Probabilistic Mechanics' since it is *a theory* of the behaviour of matter not a mathematical technique of treating experimental results. But this is not a case worth defending for two reasons: the existing name is too well-entrenched and 'Probabilistic' it hardly an attractive adjective.

Appendix C: Some 'Philosophical' Considerations

> *People are born into a society whose forms and relations seem as fixed and immutable as the overarching sky. The 'commonsense' of the time is saturated with the deafening propaganda of the status quo; but the strongest element in this propaganda is simply the fact that what exists, exists.*
>
> E. P. Thompson, *Indian Historical Review*, **3** Jan. 1978 p. 247.

SM has been burdened by lots of interpretations, some of which boldly assert that the existence of quantum theories demands a completely new approach to our views of the nature of reality and the familiar concepts of philosophy. The popular accounts of quantum mechanics are often full of this startling new philosophy; the idea that a particle can be in more than one place simultaneously is a favourite. In fact, most of the 'new' concepts are centuries, even millenia, old. They are based on the rediscovery, by physicists, of ideas due to Plato, Aristotle, Plotinus and Berkeley. A detailed discussion of these now discredited ideas is certainly not worthwhile but maybe a highly prejudiced summary might illustrate the nature of this material. There are two main branches which are worth looking at: the existence of real material objects and the possibility of *a priori* science and mathematics here treated, rather scornfully, in historical order and, finally, how these have been uncritically imported into the interpretation of SM.

C.1   Berkeley and Idealism

George Berkeley (1685–1753) was the first and most coherent advocate of the idea that nothing which one might call today 'real' exists in the world since each one of us can only think about the world in terms of our perceptions of the surroundings in his *A Treatise Concerning the Principles of Human Knowledge*; summarised as 'we can only sense our sensations.' This view is usually quoted in the phrase *esse est percipi*.[b] Berkeley was a brilliant man who wrote some of the most cogent criticism of the work of Newton on the calculus;

---

[b]'To be is to be perceived'.

all his works showed a penetrating, logical mind. The attraction of his idealistic philosophy was, of course, that it cannot be refuted by logical discussion. This possibility has proved irresistible to many. The meaning contained in 'we sense our sensations' is trivially true, its truth is contained in the meaning of the words 'sense' and 'sensation' but it is only true if all we do is 'sense' and do not 'act' on the world.

A strong, if fictional, advocate of this view is O'Brien, Winston's torturer and brainwasher in George Orwell's *1984*. In order to illustrate the invincibility of the Party he taunts Winston by saying: 'Reality is inside the skull, ... nothing exists except through human consciousness'. This view, as Orwell recognised, seems particularly appealing to those with an authoritarian cast of mind precisely because it can not be refuted by *argument* and can give the speaker a spurious feeling of omnipotence.

However, as Boswell writes in his *Life of Dr. Johnson*

> After we came out of the church, we stood talking for some time together of Bishop Berkeley's ingenious sophistry to prove the nonexistence of matter, and that every thing in the universe is merely ideal. I observed, that though we are satisfied his doctrine is not true, it is impossible to refute it. I never shall forget the alacrity with which Johnson answered, striking his foot with mighty force against a large stone, till he rebounded from it — 'I refute it thus.'

Both Boswell and Johnson recognised Berkeley's statements as *sophistries* not science. Science, like life in general, is not just logic and mathematics, it is an *investigative and experimental* activity. Logic is just a machine which can generate truths from other truths but not discover new truths. Johnson saw this immediately; the challenge to Berkeley is not through logic but action on the real world.

No one, however strongly they uphold the sophistry *esse es percipi* in their *writings*, actually believes it in their life; if their child falls seriously ill they take the child for clinical treatment developed *in practice* using new concepts and methods unsuspected by Berkeley; they are not stoically satisfied by observing the child die and be sure that nothing can be done. More importantly, anyone can conceive and even create new objects which everyone may perceive and not passively spend their lives perceiving existing 'things'; in short, all of us may *act* on the world.

There is a similar pseudo-problem associated with the idea that causation cannot be proved; all that we can say for sure is that an 'order of events' is seen. This is perfectly true if one is content to only sit in a deep armchair in a mahogany-panelled room and think about observing events. In order to investigate causation one must *act* on the world and interfere with the sequence of events. Berkeley's sophistries did not rule for long; some years after his death the static philosophy of the 18th century gave way to the evolutionary philosophy of the 19th, typified by Hegel:

> ... work, together with the transformation of the world of things, brings about the transformation of human consciousness.

C.2   The Fashion for *a priorism*

Albert Einstein once said 'the reason why we can use mathematics in physics is because physics is so simple'. He means that many of the laws of physics are capable of being expressed by a few basic axioms and this means that logic and mathematics can be applied with a confidence that no considerations which would interfere with the deduction of the structure and processes of the subject matter are likely to have been missed. By and large, basic physics is the only science (physical or otherwise) which enjoys this happy circumstance.

However, because physics is of this logically simple form it has encouraged the belief, which goes all the way back to classical Greece, that it is possible to obtain the laws of physics by thought alone, thus removing the necessity of experimental investigations. This belief has been present, together with its indignant rejection, intermittantly in western science for millenia and is still with us today; Eugen Wigner, one of the pioneers of QM, has said: 'The miracle of the appropriateness of the language of mathematics for the formulation of the laws of physics is a wonderful gift which we neither understand nor deserve'.

It is easy to appreciate the appeal of this idea since it obviously fulfils Bertrand Russell's opinion of the advantages of mathematical first principles over hard creative science:

> The method of 'postulating' what we want has many advantages; they are the same as the advantages of theft over honest toil.
>
> *Introduction to Mathematical Philosophy*, 1919

This opinion has some obvious affinity to those fundamentalists who believe in creationism, in that mathematicians look at the immense utility of mathematics in science, particularly physics, and think: 'this cannot be a coincidence it must be part of a plan'. Creationists look at the complexity of living things and their exquisite suitability to their environment and surmise: 'this cannot be a coincidence it must be part of a plan'.

In both cases what has been ignored is the importance of *evolution*; both parties are looking at a single empirical snapshot of the current state of a long and involved historical *process*. In fact mathematics, like language, is a system of formal rules for the manipulation of formal objects which has been developed, by abstraction, from millennia of practical and cultural interactions with the material world. Yet no one is amazed that the laws of science can be expressed in language. It is only if one ignores these kinds of evolutionary processes can it seem reasonable to think that mathematics, like language, now seems to be autonomous and conclude that it can exist without humans.

The idea that reality can be derived by pure thought — logic and mathematics — originates from classical Greece; Plato and Aristotle,[1] roundly and fiercely attacked by Francis Bacon, both suffered from this delusion and scorned the idea of experimental investigation as only fit for artisans and slaves. This idea recurs throughout the subsequent history of western thought; Eddington gives the most extreme view:

> 'An intelligence, unacquainted with our universe, but acquainted with the system of thought by which the human mind interprets to itself the contents of its sensory experience, should be able

to attain all the knowledge of physics that we have obtained by experiment.'[2]

A. Eddington, *Relativity Theory of Protons and Electrons*, Cambridge University Press, 1936, p. 327.

but for the purposes of QM the most famous is due to Dirac:

A good deal of my research in physics has consisted in not setting out to solve some particular problem, but simply examining mathematical equations of a kind that physicists use and trying to fit them together in an interesting way, regardless of any application that the work may have. It is simply a search for pretty mathematics. It may turn out later to have an application. Then one has good luck.

P.A.M. Dirac, *International Journal of Theoretical Physics*, **21**, 1982, p. 603.

Compare this to the first and most prominent champion of science and technology Francis Bacon, writing in 1605; here are his robust thoughts about the works of the mediaeval schoolmen:

For the wit and mind of man, if it work upon matter, which is the contemplation of the creatures of God, worketh according to the stuff and is limited thereby; but if it work upon itself, as the spider worketh his web, then it is endless and brings forth indeed cobwebs of learning, admirable for the fineness of thread and work, but of no substance or profit.

The Advancement of Learning

C.3 The Copenhagen Interpretation

The 'Copenhagen Interpretation' dates back to the very beginnings of QM and shows the influence of both idealism and *a priori* thought, albeit in a much less consistent manner than the founders of these ideas. It is, in fact, a bundle of barely coherent pronouncements the basis of which is threefold:

(1) Neils Bohr and Werner Heisenberg were fascinated by their rediscovery of Bishop George Berkeley's 18th century idealistic philosophy. Unfortunately they did not appreciate the absurd

consequences of this view. The main problem associated with this approach is obvious: it inevitably leads to solipsism; each individual believes themself to be the only thing in existence and that things cease to exist when not perceived. It is not hard to guess how Berkeley — being a bishop — 'solved' this problem. He postulated the presence of an omniscient, omnipresent, unsleeping observer who maintained the existence of the world by keeping everything under constant observation and allowed everyone to relax their attention from time to time.

(2) Certain *mathematical* similarities between the SE and the wave equation.

(3) The absence of an understanding of the theory of probability and its relationship to statistics in the days before Kolmogorov provided the basic mathematical methods on which an interpreted theory depends.

It is quite hard to say what the Copenhagen interpretation actually is since there seem to be as many flavours of it as there are defenders. It is indeed a broad church but, if one can agree to one or more of the following three 'principles' then one can claim to be a follower of this interpretation of quantum theory:

(1) The referent of quantum theory — the ultimate subject matter of the interpretation — lies somewhere in the region starting with measurement (instrument readings, for example) through the recording of such measurements to an individual becoming conscious of the measurements, generalising Berkeley's 'perception' to include instrumental measurements but, nevertheless, *esse est percipi*.

(2) The entities generating the measured quantities are capable of having particle-like behaviour or having the behaviour of waves depending on the measurement process, the experimental setup, hence the wave–particle duality.

(3) The probabilities occurring in quantum theory are 'special' in the sense that, unlike coin tosses or dice throws, they refer to individual systems not to sets, hence the deep mysteries

and paradoxes essential for a full enjoyment of the Copenhagen interpretation.

It is a little on the unfortunate side that the second of these principles contradicts the first one since, like Kant, it implicitly assumes the existence of something 'out there' generating the observations.

This doctrine, although now completely ignored by philosophers, is still seen by many physicists as 'solving the problem of the physical interpretation of QM'. The most tenacious item of this doctrine is the 'wave/particle' duality which absolves physicists from the onerous task of having to develop a realistic and non-contradictory interpretation of QM by being able to appeal to either the 'wave properties' of matter or its 'particle properties' to sidestep glaring difficulties of interpretation.

Some perspective on the effects of the Copenhagen interpretation can be gained by comparing two statements by scientists from other disciplines. In the 1950s, Hans Eysenck announced what turned out to be a futile attempt to make psychology into a quantitative science. He said that he wanted it to be 'like physics' implying that physics was the pinnacle of the exact sciences. In 2003, Vilayanur S Ramachandran, in his Reith lectures (broadcast by the BBC), discussing some inconsistencies between psychology and neuroscience, said that maybe it was not necessary to be consistent in interpretation but be 'like quantum physics[c]'. So physics from being the model of clarity for all sciences had become an excuse for sloppy thinking.

There will be no discussion of this doctrine in this work except a single remark on each of the three points given above:

(1) All physical science assumes the objective existence of the world independent of any observation, not least because the theories of physical science do not contain any mention of measurements and observations nor do the mathematical articulations of the theories contain any variables to describe their effects.

---

[c]So maybe our very notion of causation requires a radical revision here as happened in quantum physics.

(2) Mathematical similarities between the articulation of different scientific theories are very common.[3]

(3) Probabilities are real-number relative measures of subsets of a given set; these numbers *per se* contain no mention of randomness. Statistics is a set of numerical techniques to obtain optimal information from sets of measurements and may or may not involve the estimation of probabilities. In order that statistical measurements of probabilities can be reliable it is often necessary to make repeated *measurements* randomly.

It is quite pointless to enumerate the failings, contradictions[d] and absurdities of this doctrine since, like homeopathy, no amount of evidence to the contrary seems to pierce its armour. Indeed, perhaps like homeopathy, the value to believers lies in the placebo effect: assurance by an acknowledged expert and a few kind words of comfort.

---

[d]Particularly since contradictions are not seen as faults in this philosophy.

Endnotes

[1] Even Aristotle showed signs of realising the inadequacy of his stance on the *a priori* approach when he began to study biology!

[2] This view, and similar views expressed by Jeans, was thoroughly routed by Lizzie Stebbing in *Philosophy and the Physicists* Methuen, 1947. This book is still one of the best refutations of the *a priori* view, particularly if one mentally replaces 'Eddington' and 'Jeans' by the more contemporary holders of these views.

[3] A familiar simple example is Gibbs' phase rule $F = C - P + 2$ in which F is either the number of degrees of freedom, C the number of components, and P the number of phases in thermodynamic equilibrium or, Euler's rule in the same notation, F is the number of faces, C is the number of edges, and P is the number of vertices of a polygon (without holes) and all have no necessary consequences for the subject matter of those theories.

Chapter 3

The Dimensions of Space

3.1 Introduction

This chapter attempts, by means of elementary examples to reinforce and bolster the claim that the *referent* of Schrödinger's Mechanics (SM) is the set of all possible trajectories which, however natural such an assumption seemed, was simply asserted on the basis of Schrödinger's original derivation of the Schrödinger Equation (SE) starting with the Hamilton–Jacobi Equation (HJE) in Chapter 2.

The first thing to say is that the SE is a 3D[1] partial differential equation. In those cases where it can be solved exactly, the technique used is to find a set of coordinates in which the equation separates into three simultaneous ordinary differential equations *linked* by separation parameters. If this is possible then the state function is the product of the limited set of solutions of these three equations allowed by the value(s) of the separation parameters. The most familiar case is the Hydrogen atom — fortunately quantum chemistry always takes place in 3D. The opposite case is problematical; solving 1D equations and generalising to 3D is fraught with danger. If the referent of the SE is the set of all possible trajectories consistent with the energy then, obviously, this set will be very different in one, two and three dimensions and it would be astonishing if the solutions of the SEs did not reflect this difference.

The two most familiar examples of soluble 1D equations of scientific interest are:

1. The 1D free particle and its quantised analogues. Hamiltonian operator:

$$\hat{H} = \frac{\hbar^2}{2m}\frac{d^2}{dx^2}$$

together with cyclic or fixed boundary conditions to ensure Hermiticity.

2. The 1D simple harmonic oscillator. Hamiltonian operator:

$$\hat{H} = \frac{\hbar^2}{2m}\frac{d^2}{dx^2} + \frac{1}{2}k(x - x_0)^2$$

with force constant k.

These solutions often provide a basis for the computational technologies used to obtain approximations to otherwise insoluble problems.[2]

It is often said during the derivation that attention is confined to the 1D case since extension to three dimensions is straightforward or even trivial. This is not true in general but what *is* true is that extension of the formalism from one to three dimensions only *may* be trivial for the trivial 1D cases such as the two quoted above *if Cartesian coordinates are used* and even when this is possible, errors still remain. It will be shown in Section 3.3 that reducing the spatial dimensions of a problem for what seem to be obviously valid reasons in CPM does, in fact, yield wrong qualitative and quantitative results in SM. What is more interesting from the point of view of this work is the profound effect that dealing simply with a 1D case has on the *interpretation* of the results. After all, quantisation is all about constraining the motion of particles in various ways and a reduction of the space available to a particle is just such a constraint. Reducing the space available to a particular system obviously has a drastic effect on the number of trajectories available to particles in the system under study. That this problem occurs in SM but not in the HJE is due to the simple fact that the HJE is capable of generating individual trajectories when supplied with suitable initial conditions. These restricted *individual* solutions are not affected by the number of dimensions available to the motion, as the solutions of the SE are, because the referent of the SE is the set of *all* trajectories which set depends acutely on the number of dimensions available.

3.2 Paths and Trajectories

In this section the important distinction between a 'path' in space and a particle 'trajectory' is emphasised; a path is simply a continuous curve in a space and a trajectory is a path which is taken or 'occupied' by a particle. Roughly speaking, a path is a sidewalk and a trajectory is a pedestrian on a sidewalk and, of course, a single path may be the path taken by many trajectories depending on their initial conditions. Also, the interpretation of the mechanics here is, at first sight, a rather cavalier mixture of the HJE and SE interpretations but this is consistent with the assumption in Chapter 2 that the interpretation of the solutions of the SE would take over the interpretation of the HJE and Poincaré's probability interpretation of Jacobi's last multiplier of Hamilton's equations.

When we have to interpret the probability distributions generated from the state functions of the two simple cases of the last section, it seems that in both cases there is only one trajectory for the corresponding classical system (or two if, in the first case one counts both possible directions of motion). But, if one looks at the solution of the classical HJE for these two systems and its interpretation, this fallacy is exposed. Here, one obtains all the possible trajectories only when the initial conditions for each one is provided, the HJE correctly says that there are an infinity of trajectories differing in those very initial conditions which are, of course, the infinite number of initial starting points of these trajectories. What is more, these trajectories are all along the same path (the only path in each of these cases). A little thought shows that this is also the case for the probability distributions obtained by solution of the SE. For a given energy, there *are* infinitely many trajectories associated with the solution of the SE but in this case they only differ by the initial conditions of the motion in exactly the same was of those of the HJE. These individual trajectories are not recoverable from the probability distribution of SM.

> All these trajectories are along the only path available and so a *normalised* probability distribution is exactly the same whatever number of trajectories are being considered; in particular, for either one free particle or pendulum or an infinite number in these two cases.

This difference of interpretation looks like hair-splitting for the 1D systems but it becomes much clearer if the system is made more realistic by retaining the analogous Hamiltonian operator and working in a space of higher dimensions — say three, since the SE is a 3D equation — and looking at the solutions for the two cases mentioned on page 56.

Both the SE and the HJE for a free[3] particle in 3D separate in an infinite set of coordinate systems. If we ignore the freedom of fixing the origin and orientation of axes, the 11 orthogonal coordinate systems of 3D Euclidian space are enough to think about. Some of these 11 systems have an infinite number of members depending on parameters such as the eccentricity of ellipses etc. which is also ignored here. Each of these separation methods, when used to solve the HJE, generate all possible trajectories for such a particle which are, naturally, all on *straight line paths*. But these straight-line trajectories occur collected together in different 'families' according to criteria determined by the separation properties of both equations in each distinct coordinate system.

Thus, in the HJE the familiar Cartesian system generates trajectories composed of paths parallel to each axis, the spherical polar system generates families of trajectories with the same angular momentum about the origin, which amounts simply to straight lines in all possible directions which have the same perpendicular distance from that origin, and so on for the other, less familiar cases. Although the 'families' are very different the totality of possible trajectories is the same, the different families correspond to dividing up the totality in different ways.

Any one particle on one of the trajectories is a straight line and has, naturally, a constant linear momentum and a constant angular momentum; but all the members of the same linear momentum family do not have the *same* angular momentum and not all the angular momentum family members have the *same* linear momentum.

The situation in SM is *exactly* the same although the historical terminology is different. In place of the description in terms of families of trajectories for each coordinate system, we say that:

- Each family generated by the Cartesian separation has 'constant linear momentum'; the linear momentum operators commute with the Hamiltonian and the state functions have Cartesian symmetry.
- Each family generated by the Spherical Polar system has 'constant angular momentum', i.e. tangents to the same sphere; the angular momentum operators commute with the Hamiltonian and the state functions have spherical symmetry.
- For the other nine systems, each has its own set of separation operators with their associated eigenfunctions classifying the family.

Non-commutation of operators associated with separations by different coordinate systems of the SE corresponds exactly to the classification of trajectories into 'families' by separation of the HJE in different coordinate systems. None of these crucial considerations of interpretation are at all possible if one concentrates on simply mechanically 'generalising' a 1D SE. One can hardly avoid being driven to the conclusion that the origin of the conviction that the probabilities occurring in SM refer to a single system (particle trajectory, for example) is, in a very large part, due to working with very simple 1D SE where the probability distribution for a single trajectory is necessarily exactly the same as that for all possible trajectories. The sections provide some detail and following justification for the qualitative discussion above.

It might seem, since each of the 11 systems has two operators which commute with the Hamiltonian, that this implies that there are 22 possible operators commuting with the Hamiltonian. This is not the case since many of the 11 systems share at least one type of coordinate; not all are independent and, of course, do not always commute with each other. For example, the circular and elliptical cylindrical systems share a coordinate (z, say) with the Cartesian system and therefore share one of the set of operators commuting with the Hamiltonian, namely

$$-\frac{\hbar^2}{2mr^2}\frac{\partial^2}{\partial z^2} \quad \text{that is:} \quad -\frac{\hbar^2}{2I}\frac{\partial^2}{\partial z^2}$$

the angular kinetic energy (I is the moment of inertia).

3.3   Three-Dimensional SEs

Both the 1D equations of the previous section can be extended to the more realistic (i.e. *real*) 3D case which is a valuable exercise since it throws light on the physical interpretation of the SE.

Setting aside what one might call the 'trivial' differences between the abundance of 3D coordinate systems (choice of origin, orientation of axes, separation of foci — eccentricity — of ellipses in elliptical systems, etc., and confining attention to orthogonal systems), there are still 11 coordinate systems to consider; Cartesians plus 10 'curvilinear' systems. Either of the most familiar ones (to chemists, at least): spherical polars (atoms) and prolate spheroidals (diatomic molecules)) will be sufficient to show the essence of the problem.

HJE and SE are both separable in these two systems for the free particle and the isotropic harmonic oscillator; the free particle problem is, of course, separable in all 11 systems, while the harmonic oscillator, like the hydrogen atom, is separable in just four.

3.3.1   The Free Particle

The time-independent SE for any spherically symmetrical potential V including $V \equiv 0$ can be written in spherical polar coordinates (r, θ, ϕ) as

$$\left\{ -\frac{\hbar^2}{2m}\left(\frac{1}{r^2}\frac{\partial}{\partial r^2}\frac{r^2\partial}{\partial r}\right) + \frac{\hat{L}^2}{2mr^2} + V(r) \right\}\psi(r,\theta,\phi) = E\psi(r,\theta,\phi)$$

$$(3.1)$$

where $\hat{L}^2$ is the 'square of the angular momentum' operator[4] which has the familiar spherical harmonic solutions (here written $Y_{\ell,m}(\theta, \phi)$). This equation separates neatly into the equation for the spherical harmonics and the 'radial equation' may be simplified by the substitution $u(r) = rR(r)$ to

$$\left\{ -\frac{\hbar^2}{2m}\frac{d^2}{dr^2} + \frac{\hbar^2}{2mr^2}\ell(\ell+1) + V(r) \right\} u(r) = Eu(r) \qquad (3.2)$$

where the fact that the eigenvalues of $\hat{L}^2$ are $\ell(\ell+1)$ (for positive integer ℓ) has been used. Using the substitution, familiar from the 1D case

$$E = \frac{\hbar^2}{2m}k^2 \quad \text{i.e. } k = \sqrt{\frac{2mE}{\hbar^2}}$$

to introduce a more convenient variable $\rho = kr$, it is easy to show that the solutions for u and hence R when $V \equiv 0$ are the Spherical Bessel functions

$$\sqrt{\frac{\pi}{2\rho}}J_{\ell+1/2}(\rho) = j_\ell(\rho)$$

so that the (un-normalised) solutions of the SE for a free particle are

$$\psi_{k,\ell,m}(r, \theta, \phi) = j_\ell(kr)Y_{\ell,m}(\theta, \phi) \qquad (3.3)$$

where k plays the role of a continuous 'quantum number' for this energy-unquantised motion and the $Y_{\ell,m}$ are the familiar spherical harmonics.

The first thing to say about these functions is that, in view of the form of the Bessel functions, they are most definitely not eigenfunctions of the *linear* momentum operators (e.g. $\partial/\partial x$, for some choice of Cartesians) no matter in what coordinates these operators are expressed. They are in fact, by virtue of the spherical harmonics, eigenfunctions of the *angular* momentum operators. This fact is in obvious need of some interpretation. We have solved the SE not the related HJE, but it is illuminating to say what we intuitively expect from the interpretation of the description of the motion by the HJE and see how this and the comments in the earlier section bear on the solutions of the SE. This interpretation has been partly preempted in the last section but the details are interesting.

The simplest solutions are the ones with $\ell = m = 0$ so that $Y_{0,0} \equiv$ constant. These are the solutions corresponding to zero angular momentum, i.e. all the momentum is radial. That is, these trajectories are all straight lines from and towards the origin for all possible values of θ and ϕ. The (un-normalised) radial distribution of the simplest of these 's-type' orbitals is given by

$$[j_0(kr)]^2 = \left[\frac{\sin(kr)}{kr}\right]^2 = \frac{\sin^2(kr)}{k^2r^2} \tag{3.4}$$

illustrated below.

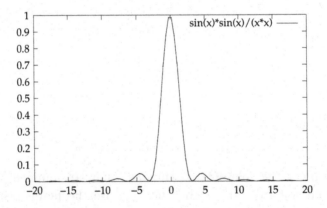

This function peaks sharply[5] at the origin since all trajectories must pass through this point and thus the probability density has its maximum here and falls off with distance from the origin as the trajectories 'fan out' to give spherical symmetry.

Other solutions with $\ell \neq 0$ correspond to trajectories not passing through the origin. Ones with the same angular momentum must be tangential to a sphere since the angular momentum of a straight-line trajectory is given by

$$L = \boldsymbol{r} \times \boldsymbol{p} = |\boldsymbol{r}||\boldsymbol{p}|\sin(\alpha)$$

where α is the angle between the r (from the origin) and p the trajectory; $|r|\sin(\alpha)$ is the radius of a sphere to which this family of trajectories are tangents.

This is rather an involved way of saying that, unlike the 1D case, there are many paths and many trajectories meaning that the 3D probability distribution is definitely not the same as that of any one trajectory!

> The fact that these eigenfunctions are not eigenfunctions of the linear momentum operators does *not* mean that the trajectories are not straight lines — they must be since they are determined ultimately by $F = \dot{p}$ — nor that their linear momentum of each trajectory is not conserved. What it *does* mean is that the whole *family* of them are straight lines but *not in the same direction*. Similarly, the trajectories described by eigenfunctions expressed in (separable) Cartesians are all straight lines which are, individually, of constant angular momentum but the whole family of them do not have the *same* angular momentum.

In summary, in the SM case we are dealing with all possible trajectories so that, in order for the operators for linear and angular momentum to commute, all the members of a family generated by separation of a given coordinate system must have the same value of *both* linear and angular momentum which is not possible.

The similarity between (1D) radial equation for $\ell = 0$:

$$-\frac{\hbar^2}{2m}\frac{d^2}{dr^2}u(r) = Eu(r)$$

and the 1D Cartesian equation:

$$-\frac{\hbar^2}{2m}\frac{d^2}{dx^2}\psi(x) = E\psi(x)$$

is misleading; the family of trajectories described by the former do not have to have initial conditions in the direction of the origin while those of the latter equation must, since there are no other directions.

3.3.2   The Harmonic Oscillator

It is not necessary to be quite so verbose in discussing the 3D isotropic — force constant independent of direction — harmonic oscillator where the choice of coordinate systems is much more restricted but the SE still obviously separates in spherical polar coordinates to give:

$$\left\{ -\frac{\hbar^2}{2m}\frac{d^2}{dr^2} + \frac{\hbar^2}{2mr^2}\ell(\ell+1) + \frac{1}{2}kr^2 \right\} u(r) = Eu(r) \qquad (3.5)$$

The unnormalised solution of the radial equation in this case is

$$R_{n_r,\ell}(r) = \frac{u(r)}{r} = r^\ell \exp\left(-\frac{1}{2}\kappa r^2\right) \times L_{\frac{1}{2}(n-\ell)}^{(\ell+\frac{1}{2})}(\kappa r^2) \qquad (3.6)$$

where

$$\kappa = \frac{k\sqrt{(m)}}{\hbar}$$

and the $L_{\frac{1}{2}(n-\ell)}^{(\ell+\frac{1}{2})}(x)$ are the generalised Laguerre polynomials.

Interestingly, the eigenvalues of the SE in this case are:

$$E_{n_r,\ell} = \hbar\sqrt{\frac{k}{m}}\left(2n_r + \ell + \frac{3}{2}\right)$$

where n_r is the 'radial' quantum number and ℓ, as usual, the angular momentum quantum number. This result generates energy levels which prove to be identical — as they should be — to the eigenvalues when the separation is done in Cartesians

$$E_{n_x,n_y,n_z} = \hbar\sqrt{\frac{k}{m}}\left(n_x + n_y + n_z + \frac{3}{2}\right)$$

in both value and degeneracy but the breakdown into radial and angular parts is more obvious and involves the angular momentum quantum number as well as a radial quantum number. Again, the differences between the two sets of *eigenfunctions* is striking; Hermite polynomials in the Cartesian case and generalised Laguerre polynomials in the polar case reflecting the different families of trajectories which are the referents of the solutions of the SE in the two cases involving the angular momentum quantum number as well as a radial quantum number.

One can obtain many more analogous results by separating the SE in any of the other 11 systems but one is sucked deeper and deeper into more exotic sets of orthogonal polynomials without obtaining any new qualitative insights into the interpretation of the solutions of the SE.

But there are *quantitative* errors in restricting the dimensionality of a problem as well as the confusions of interpretation looked at in this section.

3.4   Solutions in One, Two and Three Dimensions

One can make apparently correct simplifications of dynamical problems in quantum mechanics (QM) and get surprising results. In this section a completely soluble system, the Kepler problem — the hydrogen atom — is 'simplified' and solved by using what seems to be completely valid assumptions — one assumption from classical and one from QM. First, though, the exact 3D solution. The bound-state energies of the H atom are given by

$$E_n = -\frac{R}{2n^2} \quad \text{where } n = 1, 2, 3, 4 \qquad (3.7)$$

with degeneracy n^2 and eigenfunctions depending on the coordinate system used to separate the SE. Here, R is the Rydberg constant.

The potential is $-e^2/r$, a function of the distance from the nucleus only and therefore there are no torques acting. The classical motion for the orbiting particle is therefore in a plane whose orientation in 3D space depends on the (arbitrary) direction in which the motion starts. In CPM, this simplifies the solution by using one of the 2D systems, e.g. plane-polar coordinates (r, ϕ).

At first sight, this would seem to imply that the SE can be simplified similarly by working in the same 2D coordinates. In fact, it is perfectly possible to solve this problem. Unfortunately, the resulting bound-state eigenvalues are

$$E_n = -\frac{R}{(n + \frac{1}{2})^2} \quad \text{where } n = 1, 2, 3, 4 \tag{3.8}$$

with eigenfunctions depending on the coordinates used.

This seems a curious paradox; that which works in CPM does not work in SM. The solution is that the 2D and 3D systems have quite different sets of possible trajectories and, while the explicit trajectories may be found from the solutions of the HJE, these are not available from the solutions of the SE. In other words, we can actually specify the initial conditions for a trajectory in CPM but we are limited to the probability distribution for all possible trajectories in SM. There is no way to specify the initial conditions of a trajectory in SM and the whole set must, therefore, include trajectories in all possible planes in 3D since this is what we see under experimental conditions. The fact that there are no torques acting does not mean that there are no rotational motions, simply that there are no changes in such motions and initial conditions may involve motion in any plane whatsoever. Clearly if one could arrange to obtain a planar H atom experimentally,[6] its eigenvalues would be those given by the latter equation; quantisation is about *constrained* motion, and confining a particle to a plane is an extreme example of constraint.

Having seen the effect of a classical simplification on the SE it is worth starting from the opposite direction and considering the effect of making a similarly 'obvious' simplification on the SE itself.

As we saw on page 61, separation of the SE in spherical polar coordinates leads to the radial equation

$$\left\{ -\frac{\hbar^2}{2m}\frac{d^2}{dr^2} + \frac{\hbar^2}{2mr^2}\ell(\ell+1) - \frac{e^2}{r} \right\} u(r) = E_n u(r) \tag{3.9}$$

and the special case of $\ell = 0$ simplifies this equation to:

$$\left\{ -\frac{\hbar^2}{2m}\frac{d^2}{dr^2} + -\frac{e^2}{r} \right\} u(r) = E_n u(r) \tag{3.10}$$

and this equation is easily solved for the radial factor for the $\ell = 0$ ('s-type') state functions of the H atom.

If we go further than the last section in restricting the space available to the electron and set up the 1D 'H atom' we have

$$\left\{ -\frac{\hbar^2}{2m}\frac{d^2}{dx^2} + -\frac{e^2}{|x|} \right\} X_n(x) = e_n X_n(x) \qquad (3.11)$$

with an obvious change of notation. These two equations are the same if one restricts the 1D space to $0 \le x \le \infty$, which is the range of r in the polar case and replace $|x|$ by x. Of course, the eigenvalues of these two equations are identical

$$E_n = e_n = -\frac{R}{2n^2} \quad \text{where } n = 1, 2, 3, 4$$

which is rather embarrassing in view of the confident explanation, given in the last subsection, of the difference between the 2D and 3D results being due to the different spatial constraints placed on the motion. One would rather expect a greater difference between the 1D and 3D cases due to the more severe spatial constraint being placed on the motion. The explanation appears not in the eigenvalues but in the state functions.

The two functions $u_n(r)$ and X_n satisfy the same equation but, in the solution of the 3D case, $u_n(r)$ is not the r-dependent factor in the 3D H-atom SE. Part of the simplification of the 3D radial equation involved writing the r-dependent factor, $R_n(r)$ as

$$u_n(r) = r R_n(r)$$

so the state functions associated with eigenvalues of the same value are different and, what is more, different in a crucial way. This time, the failings involved in 'simplifying' the SE are in the state functions, and not the allowed energy eigenfunctions; something which might well be missed by concentrating on the most easily measurable quantities. The fact is that the identical eigenvalues belong to different states of the two systems.

It must be said about the 1D case that this has been a constant source of discussion for many years[7] on many (mostly mathematical) grounds, and the simple treatment given above cannot be as reliable as the 2D case.

The case of the simple harmonic operator, however, is much more reliable since there is no question of the potential having a singular point. In fact the SE can be solved for all three possible cases: the familiar 1D case quoted at the very beginning of this chapter, the 3D case in the previous section and the 2D case. In Cartesian coordinates (x, y, z) for simplicity the eigenvalues for these cases are, respectively:

$$E_{n_x} = h\nu \left(n + 1\frac{1}{2} \right)$$

$$E_{n_x, n_y} = h\nu \left(n_x + n_y + 2\frac{1}{2} \right)$$

$$E_{n_x, n_y, n_z} = h\nu \left(n_x + n_y + n_z + 3\frac{1}{2} \right)$$

$$= h\nu \left(2n_r + \ell + \frac{3}{2} \right) \quad \text{from previous section}$$

which, although perhaps obvious, is not so innocent a result as it might seem. What these results say is that any state of an oscillator described in 1D space has a different energy from that oscillator described in a 2D space *and* different again from the 3D space result for a given constant relevant sum, n, of the quantum numbers n_x, n_y, n_z, i.e. $n_x = n$ for 1D, $n_x + n_y = n$ for 2D and $n_x + n_y + n_z = n$ for 3D. Of course, this difference cannot be detected experimentally since it does not affect the distance between the levels of each description which is what generates the vibrational spectra. But it is a paradox which can be explained by the increase in the number of trajectories as the available space is increased. What is also interesting is that the 1D, 2D and 3D H atom does not show the same pattern as the oscillator; each increase in dimensionality changing the quantum number by 1/2. We await the definitive solution of this system.

3.5   An Example: Feynman's Derivation

The relation between SM and Feynman's 'derivation' of the SE consists of several parts; his two postulates in the original paper (*Rev. Mod. Phys.* **20**, (1948), p. 371) are:

[I] "If an ideal measurement is performed to determine whether a particle has a path lying in a region of spacetime, then the probability that the result will be affirmative is the absolute square of a sum of complex contributions one from each path in the region."

[II] "The paths contribute equally in magnitude, but the phase of their contribution is the classical action (in units of $\hbar$); i.e. the time integral of the Lagrangian taken along the path."

In this chapter, because of the distinction between a 'path' and a 'trajectory' in Section 3.2, Feynman's terminology, which uses 'path' where 'trajectory' is meant has been changed except in direct quotes from Feynman's paper.

These postulates are not, as we shall see, sufficient to derive his result but it is worth making a few comments of their source and content. In [I], the reference to (undefined) ideal measurements is spurious and results from the then-current Heisenberg subjective interpretation of QM; the postulate is best restated, in objective terms involving no hypothetical all-competent observer, as

[I] The probability that a particle be present in a region of spacetime is a sum of the square of the moduli of contributions from each of the trajectories.

This, of course, makes the assumption, taken from SM, that one is seeking a probability distribution and is in line with the interpretation of Section 2.5.1 that the referent of SM is the set of all possible trajectories.[8]

In [II] the key assumption is made that the relevant quantity in the evaluation of the probability distribution is

$$\exp\left(\frac{iS}{\hbar}\right)$$

described there as 'the phase of their contribution'. The use of this expression is worth a short digression since it is vital to the interpretation of any quantum theory; here is a potted history:

1926 Schrödinger

Although actually taken by Feynman from Dirac's paper (below), the expression

$$\exp\left(\frac{iS}{\hbar}\right)$$

is exactly Schrödinger's definition of the *state function* for a system; the use of 'phase' here is out of place since, as Schrödinger showed in his paper of 1926, a *complex* S generates the whole state function not simply its 'phase'. The use of 'phase' here is a ghost from the wave/particle duality era.

1933 Dirac

Dirac was very concerned about the relationship of CPM to QM especially since he was fully aware of the problems of the transition from relativistic CPM to the relativistic theory. In his 1933 paper,[9] he seems to have come to the same conclusion as Schrödinger about the importance of $\exp(iS/\hbar)$ quite independently of the fundamental 1926 publication.

1948 Feynman

Feynman was alerted to the Dirac paper by Herman Jehle and proceeded to work through the implications of this quantity as an *operator* in his astonishing discovery of a very useful technique in time-dependent QM.

Although linked to the more basic 1926 paper through the use of $\exp(iS/\hbar)$, what is omitted from Feynman's two postulates is any statement of the *law of nature*, the dynamical law, which is going to be invoked to generate the nature of the dynamics. This is, in fact, supplied, but as a supplementary condition rather than a postulate (in Equation (11) on page 372 of his paper) as a requirement on the function S for each element of the (classical) particle path:

$$S(x_{xi+1}, x_i) = \text{Min} \int_{t_i}^{t_{i+1}} L(\dot{x}(t), x(t))dt$$

which is precisely Hamilton's principle for a fragment of the 'path' which leads to the canonical equations and thence to the HJE which is Schrödinger's starting point for his clean variational derivation of the SE using complex S.

With these three pieces of information as a starting point it is difficult to see how one can avoid verifying that the assumptions would eventually lead to expressions for the probability amplitudes which agree with the SE, however different the mathematical route.

The one point relevant to this chapter which must be made is the widely-held assumption that Feynman's method provides additional evidence for the idea that the classical limit of QM as '$\hbar \to 0$' is the path of the particle. This is based of Feynman's 'derivation' that the classical *trajectory* — called the *path* in the paper — of a particle may be recovered from the limit of the theory. But this is not possible because any probability amplitude (state function) which solves the SE cannot give the path of a particle *whatever the value of* $\hbar$. The SE is a *partial* differential equation — whose solutions are functions of 3D space and time and whose referent is the set of *all possible* particle paths under the given conditions. The 'path' involved in QM is not a path of a particle but the 'path' of a state function, i.e. the space development of the state function in time; the state function does not move, it changes in time. This confusion between a particle path and a probability distribution is, in part, due to the untypical simplicity of a 1D model as has been, perhaps, over-emphasised in the earlier sections of this chapter. In the solutions of any 1D SE, there is only one path in space but an infinity of trajectories. A similar argument shows that the popular notion than as the energy quantum numbers (n, say) increase,

$$n \to \infty$$

the limit of the solutions tend to the classical trajectories. It is not possible *in principle* to obtain individual trajectories from SM any more than the result of a single toss of a die can be obtained from the probability distribution. A stronger case can be made for the limit of SM being Statistical Mechanics.

Endnotes

[1] The SE for a N-particle problem is a 'many $(3N)$ dimensional' equation but all these $3N$ variables describe the motion of N particles in real 3D space.

[2] Perturbation techniques will be considered in Chapter 5.

[3] The lack of quantisation of such a system is being ignored here for convenience.

[4] This term is the angular kinetic energy

$$\frac{1}{2I}(I\omega)^2$$

where ω is the angular velocity and $I = mr^2$ the moment of inertia; it is often described as a 'centrifugal potential'.

[5] It is interesting to recall that one of the 'approximations' to the Dirac's delta function is

$$\lim_{x \to 0} \frac{\sin^2(x)}{x^2}$$

[6] Rather a tricky exercise in microscopic flattening!

[7] The latest discussion that I can find is as recent as 2016 (R. Loudon, One-dimensional hydrogen atom, *Proc. Roy. Soc. A*, **472**).

[8] This is a very early correct interpretation of the referent of the Schrödinger Equation as the set of all possible trajectories. But it seems to have been incorrectly interpreted later as the set of all possible paths, not the set of all possible trajectories consistent with the conditions of the motion.

[9] P. A. M. Dirac, *Phys. Zeits. Sowjetunion*, **3**, 64–72.

Appendix A: Separation Operators for the Free Particle

Table A.1 lists the 11 coordinate systems in which both the HJE and the SE for a free particle separate into three (soluble) 1D ordinary differential equations. The relationship to the more common Cartesian coordinates is given in each case and, in the case of the SE, the two separation operators which commute with the Hamiltonian operator and whose eigenvalues are therefore conserved dynamical quantities. The first column in the table is the name and common notation for the three coordinates, the second is the operators which commute with the free-particle Hamiltonian operator and the third the relationship of that coordinate system to Cartesian coordinates.

It must be emphasised that in every case the HJE, when solved, generates the set of all possible *straight-line* trajectories, that is, *whatever coordinate system is used* the same infinite set of straight-line trajectories are generated. However, these trajectories are grouped into more and more exotic and unfamiliar 'families' as one proceeds down the table. The same is true of the solutions of the SE; the solutions, in whatever coordinate system they are expressed, each refer to the probability distribution of the totality of *straight-line* trajectories grouped into those very same families. In the case of the solutions of the HJE, when appropriate 'initial' conditions are provided all the trajectories can be determined, but when the SE is solved only the probability of particle presence is obtained for the whole set of trajectories.

To save space the standard notation

$$\{\hat{A}, \hat{B}\} = \hat{A}\hat{B} + \hat{B}\hat{A}$$

for the 'anti-commutator' of operators has been used in the parabolic and ellipsoidal cases.

In the case of the elliptical coordinates the values of each of the parameters (d, c and a above) generates a family of coordinate systems for which the origin is between the foci of the ellipsoids of revolution of the coordinate surfaces and foci are d, c or a distant from the origin on the same line.

Table A.1

Coordinates	Operators	Cartesians
Cartesian x, y, z	$\hat{p}_y^2, \hat{p}_z^2$	—
Cylindrical r, ϕ, z	$\hat{\ell}^2, \hat{p}_z^2$	$x = r\cos\phi$ $y = r\sin\phi$ z
Parabolic Cylindrical r, ξ, η	$\{\hat{\ell}_z, \hat{p}_y\}, \hat{p}_z^2$	$x = (\xi^2 - \eta^2)/2$ $y = \xi\eta$ z
Elliptical Cylindrical α, β, z (parameter $d > 0$)	$\{\hat{\ell}_z^2 + d^2\hat{p}_z^2\}, p_z^2$	$x = d\cosh\alpha\cos\beta$ $y = d\sinh\alpha\sinh\beta$ z
Spherical Polar r, θ, ϕ ($r > 0$)	$\hat{\ell}^2, \hat{\ell}_z^2$	$x = r\sin\theta\cos\phi$ $y = r\sin\theta\sin\phi$ $z = r\cos\theta$
Prolate Spheroidal η, α, ϕ (parameter $a > 0$)	$\hat{\ell}^2 - a^2(\hat{p}_x^2 + \hat{p}_y^2), \hat{\ell}_z^2$	$x = a\sinh\eta\sin\alpha\cos\phi$ $y = a\sinh\eta\sin\alpha\sin\phi$ $z = a\cosh\eta\cos\alpha$

(*Continued*)

Table A.1 (*Continued*)

Coordinates	Operators	Cartesians
Oblate Spheroidal η, α, ϕ (parameter $a > 0$)	$\hat{\ell}^2 + a^2(\hat{p}_x^2 + \hat{p}_y^2)$, $\hat{\ell}_z^2$	$x = a\cosh\eta\sin\alpha\cos\phi$ $y = a\cosh\eta\sin\alpha\sin\phi$ $z = a\sinh\eta\cos\alpha$
Parabolic ξ, η, ϕ	$\{\hat{\ell}_x, \hat{p}_y\} - \{\hat{\ell}_y, \hat{p}_x\}$, $\hat{\ell}_z^2$	$x = \xi\eta\cos\phi$ $y = \xi\eta\sin\phi$ $z = (\xi^2 - \eta^2)/2$
Paraboloidal α, β, γ (parameter c)	$\hat{\ell}_z^2 - c^2\hat{p}_z^2 + c(\{\hat{\ell}_y, \hat{p}_x\} + \{\hat{\ell}_x, \hat{p}_y\})$, $c(\hat{\ell}_y^2 - \hat{\ell}_x^2) + \{\hat{\ell}_y, \hat{p}_x\} - \{\hat{\ell}_x, \hat{p}_y\}$	$x = 2c\cosh\alpha\cos\beta\sinh\gamma$ $y = 2c\sinh\alpha\sin\beta\cosh\gamma$ $z = c(\cosh 2\alpha + \cos 2\beta - \cosh 2\gamma)/2$
Ellipsoidal μ, ν, ρ (parameter $a > 1$)	$\hat{p}_x^2 + a\hat{p}_y^2 + (a+1)\hat{p}_z^2 + \hat{\ell}^2$, $\hat{\ell}_y^2 + a(\hat{\ell}^2 + \hat{p}_z^2)$	$x = \sqrt{(\mu-a)(\nu-a)(\rho-a)/a(a-1)}$ $y = \sqrt{(\mu-1)(\nu-1)(\rho-)/(1-a)}$ $z = \sqrt{\mu\nu\rho/a}$
Conical r, μ, ν (parameter $1 > b > 0$)	$\hat{\ell}^2$, $a(\hat{\ell}_x^2 + b\hat{\ell}_y^2)$	$x = r\sqrt{(b\mu-1)(b\nu-1)/(1-b)}$ $y = r\sqrt{b(\mu-1)(\nu-1)/(b-1)}$ $z = r\sqrt{b\mu\nu}$

All the coordinate surfaces are figures of revolution of the conic sections (including the 'degenerate' case of straight lines which generate planes). Diagrams of the coordinate systems can be found in *Field Theory Handbook* by P. Moon and D. E. Spencer, Springer-Verlag 1961. These authors have used alternative parametrisations of some of the elliptical systems.

Chapter 4

Momenta, Operators, and a Problem

Because of the way in which Hamiltonian dynamics is formulated in phase space, momenta play a key role in the development of any system of quantum mechanics and it is, perhaps, worthwhile stressing some points in the interpretation of non-relativistic SM since, at the present time, the only path to a relativistic theory seems to be based on analogies with the Schrödinger Equation (SE). There are number of issues. Since the number of systems for which the SE is exactly soluble is very small it is necessary to be able to form the operators which are used in variational and perturbation theories which, necessarily, must be some function of coordinates and momenta.

4.1 Distributions in General

If the distribution of any dynamical variable (A say, represented by the operator $\hat{A}$) is to be examined it is important to stress that, unlike the analogue in CPM, the distribution of that quantity is composed of the *product* of two factors:

- The value of the quantity ($\hat{A}$) at the point in space, and
- The value of the particle distribution ($|\psi|^2$) at that point

both of which are functions of space. It must also be remembered that, because of the phase space of the Hamiltonian formulation, we only have the means of writing down those properties which are functions of these basic coordinates: position in space and momentum at a point in space.

We are all used to seeing the distribution of any dynamical variable $(\hat{A})$ being given by

$$\psi^* \hat{A} \psi$$

whatever the nature of the operator $\hat{A}$. Now, while this is obvious for a simple scalar operator, since it is independent of the order of the three factors in the expression, it is not so obvious for the case when, for example, $\hat{A}$ contains a differential operator. In fact, the above expression is simply *assumed* as an additional axiom rather than being understood as a necessary part of the theory in SM. The simplest case is for the key momentum operator

$$\psi^*(x)\hat{p}_x\psi(x) = \psi^*(x)\left(-i\hbar\frac{\partial}{\partial x}\right)\psi(x)$$

This expression for the *distribution* of (linear) momentum is correct but what is not correct is the assumption that the momentum operator — that operator which gives the *value* of the momentum at a particular x — is given by the familiar form

$$\hat{p}_x = -i\hbar\frac{\partial}{\partial x}$$

The actual momentum value operator comes from the combination of the HJE and Schrödinger's notation

$$\psi = \exp(S/i\hbar)$$

whence[1]:

$$\frac{\partial S}{\partial x} = -i\hbar\frac{1}{\psi}\frac{\partial \psi}{\partial x}$$

To obtain the distribution of momentum, this *value* of the momentum at a particular x-value must be multiplied by the value of the particle distribution at that point in space: $\psi^*(x)\psi(x)$

which, naturally leads to the original expression for the momentum distribution

$$\psi^*(x)\psi(x)\left(-i\hbar\frac{1}{\psi(x)}\frac{\partial\psi(x)}{\partial x}\right) = \psi^*(x)\left(-i\hbar\frac{\partial}{\partial x}\right)\psi(x)$$

and, as usual, what is an axiom in the formal theory is a simple theorem in Schrödinger's Mechanics.

Incidentally, this division of a property distribution into two factors explains the form of the kinetic energy distribution in the quantum Hamiltonian *function* (H^{SM} of page 9) which is the starting point for Schrödinger's derivation. This does not have the 'familiar' appearance; $\psi^*(Operator)\psi$.

The kinetic energy distribution is the distribution of the square of the *momentum* not square of the *distribution* of the momentum, that is, for the x component

$$\frac{1}{2m}\psi^*(x)\psi(x)\left|\left(-i\hbar\frac{1}{\psi(x)}\frac{\partial\psi(x)}{\partial x}\right)\right|^2 = \frac{1}{2m}\left|-i\hbar\frac{\partial\psi}{\partial x}\right|^2$$

and not

$$\frac{1}{2m}\left|\psi^*(x)\left(-i\hbar\frac{\partial}{\partial x}\right)\psi(x)\right|^2$$

and, most certainly not:

$$\psi*(x)\left(\frac{-\hbar^2}{2m}\frac{\partial^2}{\partial x^2}\right)\psi(x)$$

where the latter (in Cartesians only) is, if anything, the distribution of the square of the momentum *operator*.

No doubt, from an axiomatic point of view the *source* and meaning of the 'Lagrangian' used is not a matter of interest so long as the equations, in this case the SE, is generated but for physical science it is a primary concern. Once again Schrödinger's 'Lagrangian' is not just another axiom, it is the natural consequence of the central law of Schrödinger's Mechanics (SM) and its key relationship to the Hamilton–Jacobi Equation (HJE).

The original applications of SM were to the calculation of the energies of the allowed states of the hydrogen atom and the energy changes involved in transitions between these states measured by UV spectroscopy. From Bohr onwards these transitions have been pictured as occurring when an electron of *single* atom 'jumps' from one state to another. Of course, in any spectroscopic measurement, an identical process is thought to happen in trillions of atoms at once. But now we have seen that the states of SM refer to all the possible motions with the same energy so the situation is different particularly if the states are degenerate.

4.2   Momentum Distribution

We have already seen that the distribution of the components of momentum are usually given by expressions like

$$\psi^* \hat{p}_j \psi$$

where $\hat{p}_j$ is the momentum component conjugate to q^j. This expression can be examined in more detail since there are some consequences to be drawn out.

For a single particle any state function may be written as:

$$\psi(\boldsymbol{r}) = \exp\left(S'/i\hbar\right) = f(\boldsymbol{r}) \exp[ig(\boldsymbol{r})] = f \exp(ig) \quad (\text{say})$$

where $\boldsymbol{r}$ is any set of 3 spatial coordinates, $f(\boldsymbol{r})$ and $g(\boldsymbol{r})$ are real functions with boundary conditions appropriate to the case under study.

The spatial momentum density associated with the function ψ is, of course

$$\psi^*(-i\hbar\nabla)\psi = -i\hbar f\nabla f + \hbar f^2 \nabla g$$

where f^2 is the ordinary probability of presence for the particle in 3-space.

The mean value of the momentum for this state is

$$\int \psi^*(-i\hbar\nabla)\psi dV = -i\hbar \int f(\nabla f)dV + \hbar \int f^2(\nabla g)dV$$

where dV is the volume element appropriate to the coordinates $\boldsymbol{r}$.

The first term is imaginary and the second term is real. The boundary conditions imposed on ψ (and therefore f) will be the usual ones: vanishing at infinity or cyclic and this ensures that the momentum operator is Hermitian so that the first term vanishes leaving the mean value of the momentum as

$$\int \psi^*(-i\hbar\nabla)\psi dV = \hbar \int f^2(\nabla g)dV$$

the integrand of which is immediately interpretable (see Equation (2.20)) as

(Spatial momentum density at r)
$$= \text{(particle probability of presence at } r : f^2)$$
$$\times \text{(particle momentum at } r : \nabla g)$$

Showing that any contribution to the (measurable) momentum mean value is carried by what is often called the "phase" or "gauge" factor $\exp(ig)$.

The special cases are obvious:

g **constant**: The function ψ is essentially real and the mean value of momentum is zero. Obviously, because for every momentum component in a given direction, there is one of the same value in the opposite direction since any ψ describes all possible particle trajectories.

g **linear**: (in one or more of the three coordinates) ψ is an eigenfunction in the particular momentum component with eigenvalue λ (say), and, of course, mean value λ.

g **general**: ψ is not an eigenvalue of any component of momentum and its mean value is determined by the above integral.

However, for all three of these possibilities the particle probability of presence is the same

$$\psi^*\psi = f^2$$

That is, the particle density is independent of the momentum distribution.

It is obvious that the particle "density" (probability of presence) does not determine the distribution of momentum in this system and,

of course, does not determine the kinetic energy of the system; this
is a well-known problem with density functional theories. What is
more serious, perhaps, is the possibility that there are many different
energies for the same particle density. An apparent example springs
to mind immediately: *all* the states of any free particle $\exp(\boldsymbol{k}\cdot\boldsymbol{r})$ have
the same (constant) particle density, independent of their charge or
mass, so that a perfect knowledge of the particle density of a state
does not give any information at all about the nature or energy of
that state. But this is a trap; the eigenfunctions for a free particle
are not normalisable so that, in this function 'space', the momentum
operators are not Hermitian and the whole 'theory' falls.

However, what these elementary considerations *do* show is that
multiplying a state function by an arbitrary non-constant phase
factor, while not affecting the spatial probability density, does change
the momenta and (if the phase factor is time dependent) the energy
of the state.

Even though the mean values of the momentum components
are always real, the momentum *distributions* contain an imaginary
contribution from the gradient of 'the probability of presence function' f. In the time-independent case the SE is a real equation and
the boundary conditions are real so the eigenfunctions must be either
be real — with momentum mean value zero — or complex conjugates
with equal and opposite mean values. Of course, the referent of the
latter case is the division of the referent of the former case into two
halves, each half having all possible trajectories in one of the two
possible directions.

4.3   Eigenvalues of Momenta

The momentum operator $(\hat{p}_j)$ in both classical and quantum mechanics is proportional to the partial derivative with respect to the
coordinate in which the movement is being described:

$$\hat{p}_j \overset{\text{def}}{=} \frac{\partial}{\partial q^j} \qquad \text{classical}$$

$$\hat{p}_j \overset{\text{def}}{=} -i\hbar \frac{\partial}{\partial q^j} \qquad \text{quantum}$$

the difference being that the classical $(\hat{p}_j)$ operates on the function S while the quantum $(\hat{p}_j)$ operates on $\psi = \exp(S'/i\hbar)$. Since, in Schrödinger's Mechanics, we are dealing with operators it is natural to ask if there is a larger rôle for the operators $-i\hbar\partial/\partial q^j$; are they sources of partial differential equations and what is their relationship to the Schrödinger Equation?

The three components of the quantum operator are collected together as

$$\hat{p} = -i\hbar\nabla$$

What is the interpretation, if any, of the solutions (eigenvalues and eigenfunctions) of equations like

$$-i\hbar\nabla\chi_\ell = \boldsymbol{k}_\ell\chi_\ell$$

where the (vector) $\boldsymbol{k}_\ell$ is eigenvalues and the χ_ℓ are functions of space? Do they have an autonomous existence as analogues of the solution of the SE? Do the functions $|\chi|^2$ have interpretations as particle distributions involved when the momentum components are individually conserved; when, for a particular value of ℓ (to avoid double subscripts), the vector $\boldsymbol{k}$ has components $k_j, j = 1, 3$? As written above, the vectors $\boldsymbol{k}$, which have the dimensions of momentum, may take any value whatsoever; if they are possible values of momenta, they are not quantised. In other words, in the absence of boundary conditions, the operator $-i\hbar\nabla$ is not Hermitian.

The reason for this is clear, the eigenvalue equation above was not generated from the variational problem, it has been simply postulated and so there are no 'naturally occurring' boundary conditions as there are in the SE. If we restrict the motion by applied boundary conditions such as insisting that the coordinates q^j are finite or cyclic, for example, the equation above has only discrete allowed values of $\boldsymbol{k}_\ell$.

In the cyclic case, the associated eigenfunctions are products of three factors of the same form, yielding

$$N \exp(\boldsymbol{k}_j \cdot \boldsymbol{r})$$

which generate particle distributions which are *constant* in all space. This result is, of course, extremely obvious; the probability of presence in space of a particle with constant momentum is independent of position: wherever the particle is, it has the same momentum. This result is an example of the difference in SM between equations and identities; the momentum equation given above is an *identity*, it simply states what the probability distributions must be for particles with constant momenta; results which are trivially obvious and do not involve the law of motion of SM, they bring to mind, perhaps, Newton's first law rather than his second. However, it is interesting to see that equations of identical form *do* occur when the SE is solved but they occur under very different conditions.

It is also worth mentioning one crucial feature of the momentum eigenvalues which has a bearing on their interpretation, that is, a comment on the *sign* of the eigenvalues of the momentum operators is not out of place. The eigenvalues for any momentum-component operator $(\hat{p}_j)$ take both positive and negative signs for the same modulus. This does not mean that there is such a thing as a negative momentum.[2] It simply means that the momentum may be in *either direction* of the coordinate. If we take the positive sign for a linear momentum component's eigenvalue, then the motion is from right to left (say) and the corresponding negative value represents motion from left to right; with a similar condition of other types of momenta. At the moment this is all simply an analogy, it needs further concrete evidence before being accepted.

4.4   Equations and Identities

The difference between two equations of identical form which might, therefore, be thought to have similar interpretations can be illustrated and clarified by a comparison between two of the equations arising from the solution — by separation of variables — of the hydrogen-atom SE. The requirement for the satisfaction of the dynamical law is, as usual, the equality of the Hamiltonian and energy operators

$$\hat{H}\psi = \hat{E}\psi \tag{4.1}$$

where

$$\hat{H} = -\frac{\hbar^2}{2m}\nabla^2 - \frac{e^2}{r} \quad \text{and} \quad \hat{E} = i\hbar\frac{\partial}{\partial t} \qquad (4.2)$$

Since the potential $(-e^2/r)$ is time-independent, the separation into a 3D space equation and a time equation immediately yields two equations linked by a separation parameter $(E, \text{ say})$:

$$\hat{H}\psi(q^i) = E\psi(q^i) \quad \text{and} \quad \hat{E}\chi(t) = i\hbar\frac{\partial\chi(t)}{\partial t} = E\chi(t) \qquad (4.3)$$

where $\psi = \psi(q^i)\chi(t)$ and the q^i is a set of coordinates for 3D space.

The spatial equation will separate into three separate equations for four choices of orthogonal system and the commonest one to use is the spherical polar system (r, θ, ϕ) since the variable r appears explicitly

$$\psi(r, \theta, \phi) = R(r)\Theta(\theta)\Phi(\phi) \qquad (4.4)$$

The simplest of these three equations is the ϕ-component of the angular kinetic energy (classically)

$$T_\phi = \frac{1}{2I_\phi}p_\phi^2$$

which in SM involves only $\Phi(\phi)$

$$\hat{K}_\phi = \frac{\hbar^2}{2I_\phi}\frac{d^2}{d\phi^2}\Phi(\phi) = \lambda\Phi(\phi) \qquad (4.5)$$

which is easily seen to involve the square of one component of the angular momentum equation

$$-\hbar\frac{d}{d\phi}\Phi(\phi) = m\Phi(\phi) \qquad (4.6)$$

Both Equations (4.3) and (4.6) have the same form and, *if they simply stand alone*, they both have the same kind of solution; in particular both E of Equation (4.3) and m of Equation (4.6) are continuous (unquantised) and their squares are constants and, in general unnormalisable.[3] But they do not stand alone, E and $\lambda = m^2$ are separation parameters whose allowed values are also influenced by the fact that they occur in other equations which *do* have boundary conditions imposed by the original variational problem:

- The values of E are determined by the spatial part of the separated equation and must, therefore, take *these* allowed eigenvalues not the continuum suggested by the simple time equation.
- The allowed range of values of m (and λ) are determined by the r and θ equations and the range of the separation coordinate, $0 \leq \phi \leq 2\pi$, determines the boundary conditions on $\Phi(\phi)$ leading to a limited range of integer values[4] for m.

The point of this rather long digression is to show, by example, that *unless the quantities momentum and energy are those which satisfy the equation of motion* ("$\hat{H}\psi = \hat{E}\psi$") there is no necessary relationship amongst them, they are four *independent* variables as reflected in the continuous nature of their eigenvalues unless such a Schrödinger Equation determines their inter-relationship.

In the particularly simple cases of energy and momentum (components), the physical interpretation of the squared eigenfunctions is a trivial tautology:

- The time distribution of a time-independent quantity is constant, it does not change with time.
- The spatial distribution of a constant momentum component is constant, it does not change as it moves.

This example highlights the error in assuming, as many axiomatic formulations of QM do, that there is nothing unique about the Hamiltonian operator and the associated SE; it is assumed that, for any operator ($\hat{A}$, say) 'representing' a dynamical variable (A, say) there is an equation as follows:

$$\hat{A}\Theta_i = a_i\Theta_i \qquad (4.7)$$

which determines the allowed values of A — the a_i — and the spatial distributions — the $|\Theta_i|^2$ associated with the a_i.

But this is not possible *in general* because the only allowed motions and particle distributions must be ones which obey the basic law of Schrödinger's Mechanics. Thus, in order that the above equation have solutions which involve real allowed motions those solutions must also solve some Schrödinger Equation or be a factor of a function which solves some SE. In the time-independent case, this simply boils down to the fact that any $\hat{A}$ must commute with some $\hat{H}$. That is, $\hat{A}$ must be involved in some $\hat{H}$.

It is usual also to insist that $\hat{A}$, representing A, is Hermitian, but this is simply an empty hope since it is not possible *a priori* to say whether or not an operator is Hermitian without a knowledge of the functions on which it operates without which the 'Hermiticity' cannot be shown. In particular, this involves a knowledge of the applicable boundary conditions as we have seen in the simplest possible genuine operator $\hat{p}$. If Equation (4.7) is not derived from some physical variation principle it is not easy to see where these boundary conditions would come from.[5] When setting up operators $\hat{A}$, or even when trying to form a Hamiltonian operator there are difficulties associated with the fact that there are often ambiguities of form in some in the terms involved because of non-commutation of the factors involved. This is entirely due to the fact that operators are being set up empirically and are not being formed as a Euler–Lagrange equation arising from the minimisation of some functional. Such a functional can only be found if the variable A is compatible, either explicitly or implicitly, with Schrödinger's dynamical law.

There is a related problem; it is usually claimed that the distribution in space of the variable A is of the form:

$$\psi^* \hat{A} \psi$$

with corresponding mean value:

$$\int_a^b \psi^* \hat{A} \psi \, dV$$

But it has been shown, by example, that this is not the case, or rather is only the case if $\hat{A}$ is a function of coordinates alone or is a function of coordinates and is *linear* in momenta. The kinetic energy distribution is given by

$$\frac{\hbar^2}{2m}|\nabla\psi|^2$$

and used by Schrödinger in his Lagrangian. All the distribution of dynamical variables is given by the generalisation of the verbal expression in Section 4.2 by expressing the operator $\hat{A}$ in terms of the coordinates q^j and the momentum operators

$$\hat{p^j} = \frac{-i\hbar}{\psi}\,\frac{\partial}{\partial q^j}$$

which only gives the result $\psi^*\hat{A}\psi$ for expressions linear in the p_j. Kinetic energy is, of course quadratic in the pj, hence the different result.

There is no error corresponding to Equation (4.7) made in the formulation of CPM; no one would claim that $F = \dot{p}$ or the HJE were just one of many similar equations with a claim to describing the allowed motions of particles. This is an error of false abstraction; in the Hilbert space abstract formulation of QM the Hamiltonian is just one operator with no *mathematical* claim to uniqueness; its uniqueness lies in the *physical* fact that it alone describes the motions involved in its area of applicability, the real micro-world.

4.5   A Special Case: Angular Momentum

In everything which has been discussed so far the terms 'coordinates' and 'momenta' have been used rather informally insofar as we have relied on intuition and ordinary practice to supply a picture of what is meant by a coordinate, and the definition is:

$$p_i = \frac{\partial L}{\partial \dot{q}^i} \tag{4.8}$$

to provide a conjugate momentum component; working in CPM here, L is the classical Lagrangian. The simplest examples show that this usage seems justified; in particular the familiar results in Cartesian coordinates are all consistent with this general theory. In normal practice 'coordinate' at its most complicated usually means a member of one of the familiar 11 orthogonal coordinate systems in 3D space. If one of these arbitrary 3D coordinate systems is used, then Equation (4.8) provides the conjugate momentum components which are, however, intuitively less accessible.

Quite independent of the Lagrangian and Hamiltonian formalisms, there are existing definitions of various types of momenta and so it is natural to inquire about the relationship between what one might call 'naturally occurring' momenta and their components and the momentum components conjugate to sets of coordinates. That is, under what conditions is a coordinate suitable for use in the canonical formalism and under what conditions can a 'pre-existing' momentum component be made conjugate to some coordinate in the canonical formalism?

Elementary considerations are enough to show that there is a whole *class* of momentum components which cannot be brought into the canonical formalism in spite of their utility and familiarity.

In elementary (vectorial) mechanics one defines the angular momentum *vector* as

$$\vec{\ell} = \vec{r} \times \vec{p} \tag{4.9}$$

and this coordinate-free definition implies that the angular momentum vector can be resolved into components in any coordinate system whether or not any member of the coordinate system is an angle. That is, Equation (4.9) is not concerned with the idea of momentum components conjugate to coordinates with the dimensions of angle (i.e. no dimensions) it is simply *called* the angular momentum for intuitively justifiable reasons. In fact, the most usual form of resolution of $\vec{\ell}$ is

$$\vec{\ell} = \ell_x \vec{i} + \ell_y \vec{j} + \ell_z \vec{k}$$

in Cartesian components and, of course $p_x = \partial L / \partial \dot{x}$, not ℓ_x, is conjugate to x.

What is very clear is that $\vec{\ell}$ cannot be expressed in a way in which, to each of three linearly independent components (in some coordinate system), there is a conjugate angular coordinate. This is trivially true simply because the position of a point in space cannot be specified by three angles: at least one length is required. One might think of this simple example as suggesting a kind of 'inverse problem in canonical coordinates' : given a set of momentum components, under what conditions can we find conjugate coordinates which

(1) are a complete and non-redundant set for the problem in hand,
(2) regenerate the given momentum components *via* Equation (4.8).

Now among the familiar orthogonal coordinate systems there are some which have two angles and one length as their dimensions. But the problem with the resolution of the angular momentum vector is more acute than we have suggested: as we shall see shortly, it is possible to make *only one* of the three components of $\vec{\ell}$ into a canonically conjugate momentum. Perhaps the most direct explanation of why this is so is by way of an explicit example: the transformation between Cartesian and spherical polar coordinates.

Taking

$$\vec{p} = m\vec{v} = m\dot{\vec{r}}$$

we have

$$\vec{\ell} = \vec{r} \times \vec{p} = m(\vec{r} \times \dot{\vec{r}})$$

and the Cartesian components of $\vec{\ell}$ are

$$\ell_x = m(y\dot{z} - z\dot{y})$$
$$\ell_y = m(z\dot{x} - x\dot{z})$$
$$\ell_z = m(x\dot{y} - y\dot{x})$$

using

$$x = r\sin\theta\cos\phi$$
$$y = r\sin\theta\sin\phi$$
$$z = r\cos\theta$$

gives

$$\ell_x = -I(\dot\theta\sin\phi + \dot\phi\cos\theta\sin\theta\cos\phi)$$

$$\ell_y = I(\dot\theta\cos\phi - \dot\phi\cos\theta\sin\theta\sin\phi)$$

$$\ell_z = I_\phi\dot\phi = (I\sin^2\theta)\dot\phi$$

where $I = mr^2$

where I_ϕ is the moment of inertia of a particle about the z axis. Now for a Lagrangian of the simple form

$$L = \frac{1}{2}mv^2 - V = \frac{1}{2}m(\dot r^2 + r^2\dot\theta^2 + r^2\sin^2\theta\dot\phi^2) - V$$

$$\frac{\partial L}{\partial\dot\phi}$$

is indeed $I_\phi\dot\phi$ and so the angular variable ϕ and the angular momentum component

$$p_\phi = I_\phi\dot\phi$$

are indeed conjugate. But ℓ_z is the *z-component* of the original angular momentum not the ϕ -component!

Of course, ϕ is 'tied' to the choice of z-direction (it is an angle of 0 to 2π around the z-axis) and we could have defined ϕ in an analogous way around the x-axis and so obtained $\ell_x = I_\phi\dot\phi$. But we cannot do *both* if only for the simple reason that angles of 0 to 2π around two mutually perpendicular axes cover the sphere *twice* making such a putative coordinate system redundant and, incidentally, showing that there is no possibility of a non-redundant set of coordinates containing *two*, let alone three, canonical angular momentum components. In fact, once one component of the angular momentum 'vector' has been chosen as conjugate to an angular variable this choice excludes other angular momentum components from being conjugate to any other angular coordinate in any coordinate system.

Some further insight into the problem may be obtained by considering ignorable coordinates — the starting point of the transformation theory. In Cartesians, if

$$\frac{\partial L}{\partial x} = \frac{\partial L}{\partial y} = \frac{\partial L}{\partial z} = 0$$

this usually implies the potential is a constant (say zero) and the vanishing of the above derivatives implies (via the Lagrange equations) the constancy of the three-conjugate momentum components and we interpret this constancy by saying that in the absence of a potential function, 3-dimensional space is isotropic with respect to linear displacement. Now, in spherical polars a free-particle Lagrangian is

$$L = \frac{1}{2}m(\dot{r}^2 + r^2\dot{\theta}^2 + r^2\sin^2\theta\dot{\phi}^2)$$

and

$$\frac{\partial L}{\partial \phi} = 0$$

but

$$\frac{\partial L}{\partial \theta} \neq 0$$

indicating a major a symmetry between the two angles in the coordinate system: ϕ has a privileged position; θ has the range 0 to π. Once a given axis is chosen then the homogeneity of space for *rotations* is destroyed *by that choice*. Thus, caution is required if there is a tendency to make too much of certain apparent equivalences between translations and rotations: between angular and linear coordinates.

We have seen above that p_θ is not constant for a free particle but p_θ is conjugate to the angular coordinate θ and the canonical equations can be expressed in spherical polar coordinates. It should be clear now what the source of the confusions about angular momentum actually are: they are verbal rather than essential.

We have been discussing two quite distinct concepts and confusing them together because of similar terminology and certain intuitive expectations. These confusions are compounded by a *contingent* connection between the two which occurs in a familiar coordinate system. The clues to the resolution of the confusion lie in the fact that it is the *z-component* of $\vec{\ell}$ which is conjugate to ϕ (not z) and the fact that p_θ *is* a valid momentum component conjugate to θ. The problem is simply the *simultaneous existence* of two quite different quantities: the angular momentum 'vector' and the momentum components conjugate to angular coordinates. These two quantities are defined independently of each other and, in general there will be no simple connection between them.

In fact, as we have seen, there may be a *contingent* connection between them in the sense that (in spherical polar coordinates, at least) one of the *Cartesian* components of the angular momentum vector is identical to a momentum component conjugate to an angular variable. This is nothing more or less than a coincidence which has, unfortunately, served to muddy the distinction between the two separate quantities. If one considers the 11 orthogonal coordinate systems in 3D space — many of which have dimensionless (angular) members — the role of θ in spherical polars is more typical. The momentum component $\partial L / \partial \dot{\theta}$ is a proper conjugate momentum component: conjugate, that is, to an angular variable but it is not a component of the angular momentum 'vector', in particular it is not a Cartesian component of the angular momentum 'vector'. It is always possible *mathematically* to express a vector[6] as a sum of components parallel to some axes but, at best, only one of these components can exist in reality. In short, although one may express the (Cartesian) vector into three components (ℓ_x, ℓ_y, ℓ_z) it is not possible for all these three independent components to be found experimentally.

Even if the position of a point cannot be specified by three angles it might be thought that the three Euler angles (for example), which are used to specify the orientation of a rotating body might, pass muster as 'rotational canonical' coordinates. These three angles are, in fact, not sufficient to define the orientation of a rigid body with respect to a fixed 'global' coordinate frame because one must specify the *order* in which the rotations are performed in fixing the body's orientation. Such a set of coordinates cannot be brought into the canonical formalism.[7]

> This extended discussion of some of the properties and peculiarities of angular momentum was initiated by the more general question 'under what conditions is a coordinate (or momentum component) suitable to be used in the canonical equations'. However, it is worth remarking that these confusions between angular momenta and canonical momenta conjugate to angular coordinates, when taken over into quantum theory, are at the heart of the Einstein–Podolsky–Rosen (EPR) paradox since the mistake that there can, simultaneously, exist in reality more than one component of angular momentum is crucial to this 'paradox'.

4.6   A Problem

We have seen what the *referent* of SM is but this is only a fraction of the attempt at a physical interpretation. The referent of Newtonian mechanics is particles subject to forces but this tells us nothing about the actual behaviour of such particles. What is needed is some idea of the behaviour of the referent of the SE: sets of all the possible trajectories of particles.

In the axiomatic approach to a mathematically articulated physical theory, the interpretation of the mathematical formalism, if one is given, is just another set of axioms albeit, perhaps, guided by comparisons between experiment and the results of the application of the mathematics. In quantum theories, however, there is enormous freedom in the choice of physical interpretation of the mathematical entities since comparison between theory and experiment often only involves the interpretation of a tiny fraction of the mathematical expressions involved. A strange hybrid has been developed composed on the one hand of operationalism which has concentrated on excellent numerical agreement between computed *energies* and the experimental results and, its direct opposite, a species of wild speculation of unobservable schemes have been invented, from theories of measurement to multiple universes. The major culprit escaping from the net of all of both operationalism and unverifiable speculation is, of course, the state ('wave') function.

Instead of trying to eliminate the physical interpretation of this function, which would leave very little of value left when applying SM to chemistry and electron physics where *processes* and *mechanisms are sought*, by far the best approach is to look carefully at some concrete 'simple', but nevertheless real, examples and use any method available to seek an interpretation. It therefore seems necessary, as well as attempting to derive a realistic interpretation of SM by careful comparison with its nearest relative CPM, to look more carefully at some of the internal structures generated by the mathematics and try to get some sense from these quantities. In doing this it is necessary to give a warning that the material in this chapter must seem unfashionably pedestrian; it involves a detailed inspection of the state functions of these very simple but realistic examples, which is more the interpretation of the state functions and their gradients studied at length to try to get the clearest interpretation, initially of the contributions to the kinetic and potential energy distributions of a single-particle system and hence the details of the physical interpretation of the referent of SM.

In Chapter 2, it has been shown that Schrödinger's generation of the SE can only be derived from the statement (axiom?) that the basic law of QM is that the difference between the mean value of the quantum Hamiltonian *function* distribution in space and the mean value of the spatial energy distribution be a minimum:

$$\langle H^{SM} \rangle - \langle E \rangle = \text{minimum}$$

This new law replaces the classical (HJE) law $H^{CPM} = E$, the equality of the classical Hamiltonian and the energy, where H^{SM} and H^{CPM} are defined on page 29.

This has led naturally to the interpretation of the referent of the state functions of SM as being related to the referent of the solutions of the HJE of CPM since both are functions of the same set of spatial variables and, possibly, time. Neither are functions of the position(s) of particle(s). For example, in the time-independent case, the referent of both the solution S of the HJE and ψ of the SE is a set of all allowed spatial trajectories of a given energy (in SM eigenvalue of the SE) because all points in the available space are on some trajectory.

However, when these two laws are seen side by side as above it is clear that there is to be one obvious and striking difference between the interpretation of the solutions of the HJE and the SE.

> Both statements are clearly true for the solutions of the HJE for which $H^{CPM} = E$. If, however, the equality of the *mean values* are used in place of the classical one-to-one equality, surely there must be cases (points in space) for which the distribution of H^{SM} is *not* equal to the distribution of E in order that these *distributions* be non-trivial. And we know the squares of the solutions of the SE are in fact typically distributions over all of configuration space.

The key difference between the two cases is that the classical case refers to a *single, unique* value of the system's energy plus the fact that a knowledge of S plus initial conditions gives all the possible trajectories of the particles. However, the general quantum case refers to a *mean value* of a spatial distribution of energy, and Hamiltonian function tells us nothing at all about individual trajectories; it does not, however, deny the existence of such trajectories. However, in the restricted time-independent case, the energy does have a single fixed value which leaves just a single distribution to think about: the distribution of the Hamiltonian *function*.

This, of course, means that (again in the simpler, time-independent case) the solutions of the SE while, on the one hand being functions of space and time, cannot have the referent set of all trajectories *having the value* of the Hamiltonian function when these trajectories do have the same value of the *energy*. The referent of the solutions of the SE must therefore actually be the set of all trajectories which yield that particular mean value which is generated by the solution of the SE. As will become obvious, the necessity of this new referent explains, among other things, the fact, mentioned earlier, that the kinetic energy distribution is $\hbar^2|\nabla\Psi|^2/2m$ not $-\hbar^2\Psi^*\nabla^2\Psi/2m$; for, if the ∇^2 form is used for the kinetic energy this problem goes away — all points in space are associated with the Hamiltonian distribution — but with the unfortunate and necessary consequence that, in some parts of space, the 'kinetic energy' distribution is *negative* or otherwise counter-intuitive.

More importantly, one cannot insist that, in spite of its intuitive failures, the kinetic energy must be given by the ∇^2 expressions because this definition is in conflict with that of the variational expression ('Lagrangian') involving the quantum Hamiltonian function (H^{SM}). For example, in his study of the forces and energy changes on chemical bond formation, Klaus Ruedenberg[8] used only the expression $|\nabla\Psi|^2/2m$ for the kinetic energy since it is clear that the $\Psi^*\nabla^2\Psi/2m$ is meaningless as an interpretational tool.

In earlier chapters, the continuity relationship of the CPM HJE and the SE has been emphasised but, in retrospect, this is just an historical development, and does not include an identity of interpretation. It should have been expected that *something* in the interpretation of SM must be different from that of the HJE, otherwise one could simply use the HJE with the function $S(q^j)$ unchanged (i.e. not $\psi = \exp -iS/\hbar$ in Schrödinger's variation principle) and SM would just be Classical Particle Mechanics (or, more correctly, classical statistical mechanics) in fancy dress. This calls for the examination of a particular case to see how this new requirement appears in practice.

4.6.1 The Simplest Non-Trivial Case: Harmonic Oscillator

The easiest case to consider is the 3D homogeneous harmonic oscillator since the SE separates (among other systems) the Cartesian coordinates into three 1D equations of identical form[9]:

$$\left\{\frac{1}{2m}\frac{d^2}{x^2} + kx^2\right\}\psi_{n_x}(x) = E_{n_x}\psi_{n_x}(x)$$

The solutions are Hermite Polynomials multiplied by the factor $\exp(-\alpha x^2)$ where:

$$\alpha = \frac{m\sqrt{k}}{\hbar}$$

Maybe the most obvious thing to say about the difference between SM and CPM is the fact that, while the solutions of the SE are functions which are non-zero everywhere in space — going asymptotically to zero at infinity — the motion of the classical SHM has *limits* for a given energy E say, the oscillator does not move beyond the points x for which $E = kx^2$, where the potential energy is equal to the total energy where the kinetic energy is zero as the particle reverses direction.

The lowest energy state in the classical case is, obviously, where the particle of mass m is stationary at the origin. This is a 'statics' problem not a dynamical one and so there is no HJE or, rather, the HJE is just a trivial case. The classical 'distribution function' in this case is a Dirac delta function at $x = 0$ and the system has kinetic and potential energies, both of which are zero.

The ground state of the quantum case has an energy eigenvalue of 1/2 (not zero) and a state function which is a Gaussian bell curve symmetrical about the origin (the Hermite polynomial is just 1)

$$\psi_0(x) \propto \exp(-\alpha x^2)$$

Obviously, the greater the 'stiffness' of the force field or the mass of the particle — the harder it is to move the particle — the slimmer and higher the distribution $|\psi_0(x)|^2$; the more $|\psi_0(x)|^2$ looks like the classical Dirac distribution. But the width of the distribution is never zero, that is there is always a non-zero probability for the particle to be some distance from the origin — the source of the potential. The question is, therefore, 'what is the energy of the particle at these points in space?' since it cannot be zero. It is subject to a force and so must be moving, hence have kinetic energy and it must have potential energy due to the same force field. Further, both these quantities are *positive* and so cannot cancel. These terms need individual, concrete comments. The general points of the interpretation can be obtained simply by looking at the qualitative features of the first few non-ground-state functions which do refer to actual moving (vibrating) systems.

Firstly, for the ground state, *at the origin* $x = 0$ the momentum of the particle is zero (gradient at the top of Bell Curve zero) so it is stationary and the kinetic energy is zero. Second, it is by assumption at the origin and so its potential energy is zero. The value of the distribution of energy at the origin is zero. This result hinges on assuming that the kinetic energy distribution is, in fact, $\hbar^2|\nabla\psi_0(0)/(2m)|^2$ as it is assumed in the derivation of the SE. However, if we take the value of the kinetic energy distribution at $x = 0$ from the 'kinetic energy operator' in the SE $(-\hbar^2\psi_0(0)\nabla^2\psi_0(0)/(2m))$ it is obvious from the eigenvalue equation that this is $(1/2)|\psi_0(0)|^2$ because the potential energy at the origin is zero.[a] So, the phenomena are saved (as Ptolomy would have said) but at the expense of giving a stationary particle non-zero kinetic energy: the celebrated zero-point-energy. A few moments' consideration of the simple examples above shows that the behaviour of $|\nabla\psi(x)|^2$ is, in general, quite different to that of $\psi(x)(\nabla^2)\psi(x)$ even to the extent that, in particular cases, where the former has a local maximum, the latter is zero.

There are more spectacular parts of the space which are even more counter-intuitive. Since there are no limits imposed on the value of the displacement of the motion except the limit $\psi \to 0$ as $x \to \infty$ the potential energy diverges: $\psi V\psi \to +\infty$ with the (positive) kinetic energy $|-\hbar\nabla\psi|^2/2m$ simply falling away. This may well be an artefact of *not* applying the intuitive boundary conditions for the problem as is done, for example for the 'particle in a box'.

Some comfort from this dilemma may be found in that, in the domain of applicability of SM — sub-atomic, super-nuclear particles in potential fields — there are no harmonic force fields, there are only the combined effect of various electromagnetic fields. This fact has been noted in *Probability and Schrödinger's Mechanics* as solving the zero-point energy 'conundrum'. It is also relevant here; the harmonic force field in, say, diatomic molecules is just a fairly accurate description of the motions of the nuclei for small displacements of those nuclei from their equilibrium positions. The fact is, of course that the bonds in all diatomic molecules can be broken and the nuclei cannot be forced together beyond a certain distance. In a typical

---

[a]Again, because ψ_0 is an eigenfunction the energy at all points in space is $1/2$: the eigenvalue.

diatomic molecule, say carbon monoxide (CO), the 'harmonic field' is simply the combined effect of the electrostatic interactions among the two nuclei and the 14 electrons. That one resultant of these $(16/2)(16+1) = 136$ interactions among the particles is experienced as close to a harmonic field looks, at first sight, to be rather surprising and fortuitous. But as is well known from Classical Particle Mechanics that any sufficiently small deviation from a position of equilibrium of a set of interacting particles due to *any* combination of conservative forces gives rise to a net harmonic restoring force.[10]

Obviously, interesting as this is, it is scarcely an explanation of the phenomenon; it has the feel of a 'just so story'; an *ad hoc* after-the-fact justification. Equally obviously, the 'explanation' has simply pushed the real explanation one step further back since the same thing happens when electromagnetic fields are used explicitly; the solutions of the Kepler problem (the hydrogen atom in SM), for example, shows exactly the same behaviour. In both cases, the 'potential energy' increases without limit in certain spatial regions while the 'kinetic energy' (as given by $(1/2m)|\nabla\psi|^2$) is always positive, adding to the PE in the harmonic case and subtracting from the PE in the Kepler case.

So, for the present, since, as far as SM is concerned, electromagnetic forces are fundamental, and cannot be explained away as a composite of more 'fundamental' forces this problem has to be left hanging in the air. But, as we have seen in reducing the harmonic case, the problem will still remain *whatever* the 'true' force fields are; it is an essential part of SM. We simply have to accept that a major consequence of the fundamental principle of SM — the equality *in the mean*[11] of the quantum Hamiltonian (spatial) distribution and the energy distribution — Schrödinger's 'Lagrangian' starting point from which the SE is derived — leads to the existence of regions of space in which the sum of the kinetic energy and potential energy distributions is not equal to the energy distribution of the system.

In hindsight, the existence of this problem *should* not be surprising because, if the Hamiltonian distribution were to be equal to the energy distribution at all points in space, the SE would simply be the HJE in different notation: '$\psi = \exp(S/\hbar)$'. It is the introduction of the Lagrangian of Schrödinger which generates quantum mechanics. With this fact in mind, the 'interpretation' given above of the harmonic force field not being fundamental explains the non-existence of 'zero-point-energy' but is of no help in the general case since, as we have seen, this problem of interpretation is fundamental to Schrödinger's Mechanics.

It may be worth remarking that this result is not *per se* the result of *interpreting* Schrödinger's Lagrangian as the minimisation of the mean value of the two densities, it results even if the Schrödinger Equation is simply assumed as a first principle (an axiom) with no particular interpretation given to the integrand in the Lagrangian. The result simply hinges on the expression for the kinetic energy being $(1/2m)|\hbar \nabla \psi|^2$. In fact, only by abandoning the consistency of using this last expression in the Lagrangian and using the *same* expression in interpreting the results of solving the SE can some form of *ad hoc* 'solution' to this problem be found. If one looks simply at the Hamiltonian *operator* one sees two terms, one of which is clearly the potential energy, then it is tempting to simply insist that the other one *must* be the kinetic energy, i.e. the kinetic energy must be the distribution $-\psi^*(\hbar^2/2m)\nabla^2\psi$ in which case the total energy distribution would be correct as

$$-\psi^*(\hbar^2/2m)\nabla^2\psi + V|\psi|^2 = E|\psi|^2$$

but at the considerable penalty of enduring the paradox that the existence of regions of space where the putative kinetic energy is *negative* as well as the inconsistency of having two different definitions of kinetic energy. As Ptolomy would, no doubt, have said "the phenomena are saved" but at what cost?[12]

It cannot be overemphasised that *the* law of nature in Classical Particle Mechanics (Newton's law) *is* that the condition for actual motion is that the (classical) Hamiltonian function be equal to the energy of the system at every point in space while the law of nature in the quantum case is the equality of the *means* of the (quantum) spatial Hamiltonian distribution and the spatial energy distribution. The two laws are quite distinct but obviously related; there are probabilities in the quantum case but not in the classical case.

Endnotes

[1] The fact that the momentum-value operator contains a division by ψ might cause alarm since this would obviously cause the value of the momentum to diverge at nodes in the state function; points at which $\psi = 0$. Even this apparently awkward fact can be given a realistic application. A node in the state function means there is zero probability of a particle being at that particular point. In classical terms the distribution of a particle would obviously be related (inversely) to its velocity; the faster the particle moves at any point the smaller the probability that the particle will be at that point — where the momentum is a maximum, the probability will be a minimum.

[2] Where a moving body could impact a stationary one and cause it to move in a direction opposite to its motion, perhaps.

[3] That is, strictly the operators $\hat{E}$ and $\hat{K}_\phi$ are not Hermitian since the class of functions on which they operate has not been specified.

[4] In the first edition of his famous book *Quantum Mechanics* Dirac missed this last point and wrongly said that m is continuous.

[5] In much of the applications of electron physics it is assumed that the boundary conditions are 'in our gift', they can be applied to suit the needs of the case. But this is not the case, the boundary conditions arise from the optimisation of the Lagrangian.

[6] In fact, angular momentum is not a vector but a bi-vector or antisymmetric second-rank tensor, a fact which is expressed much more clearly using coordinate-free geometric algebra. It is only the coincidence in 3D space that $3 = 3(3-1)/2$ which enables the components of this bivector to be put into one-one correspondence with the components of a vector.

[7] It is easy to show that

$$[\ell_x, \ell^2] = [\ell_y, \ell^2] = [\ell_z, \ell^2] = 0$$

but this does not admit ℓ^2 into the canonical scheme because there is no coordinate to which ℓ^2 is conjugate.

[8] M. J. Feinberg, K. Ruedenberg and Ernest L. Mehler. *Advances in Quantum Chemistry*, 1972.

[9] This example is chosen because it avoids the pitfalls of generalising from a 1D case to three dimensions discussed in Chapter 3.

[10] ∇V is zero at equilibrium and the largest term is therefore terms of the type $(\partial^2 V/\partial q^2) \times (q)^2$ which are just of the form $k\times$ (displacement)^2.

[11] This phenomenon is familiar in that any function obtained by the minimisation of a distribution cannot provide reliable results at particular individual points; fitting a square wave by a trigonmetrical series can give excellent results except at the square corners.

[12] In the solution of the variational problem it is precisely the 'replacement' of the expession $|\hbar\nabla\psi|^2$ by $-\psi^*(\hbar^2/2m)\nabla^2\psi$ which generates the Euler–Lagrange equation — in this case the SE — which ensures that the difference between the two means is minimised. As we pointed out in Section 4.1, the integrand of the Lagrangian has a physical interpretation as a law of nature but the terms in the Euler–Lagrange equation which (along with the boundary conditions) generates the solutions to the variational problem may well not have a physical interpretation.

Chapter 5

Approximation and Interpretation

The number of Schrödinger's Equations (SEs) which can be solved exactly is very small; they have all been mentioned earlier in this work (except for the fixed-nuclei hydrogen molecule cation H_2^+) and they are all essentially for a single particle in a field of force. This necessarily means that approximation methods for the investigation of more interesting systems must play a huge part in all but the simplest applications of SM. There are two general classes of approach both of which involve expanding the unknown solution as a linear expansion of some known functions. Both of them were pioneered by Schrödinger in 1926.

It is well known that the infinite number of solutions of certain types of 3D partial differential equations form a 'complete set' and are said to span the space of functions in that 3D space which has the same boundary conditions. Further, linear combinations of products of n members of any complete set will span the space of dimension $n3D$ satisfying the same conditions. If, therefore, the forms of the functions are known, then all that is necessary is to calculate the coefficients in the linear expansion of a solution of an otherwise intractable SE. That is to say an approximation to a solution of a SE can (in principle) always be found by using a finite number of known functions and calculating the coefficients associated with a particular choice of expansion functions; replacing the problem of solving a partial differential equation in a many-dimensional space (a problem in mathematical analysis) by an algebraic problem. Since algebra is always easier than analysis, these considerations provide the basis for all approximation methods in SM. Obviously, what is required before this simplification can be carried out is a method for determining the linear expansion coefficients; a criterion which the linear expansion must satisfy.

The power of this approach can be seen, for example, by the fact that the calculation of an approximate energy and state function for a very simple molecule like water is simplified from being tables of billions of values of the state function in space to a few dozen expansion coefficients. A concrete example of the magnitude of the saving involved is given by dispensing with expansion altogether and solving the problem numerically as Hartree and his father did for atoms in the early days — without computers. In fact, Hartree junior once said that it is impossible to calculate a state function of the ground state of the uranium atom (92 electrons) since, even if all the matter in the universe were turned into paper, there would still not be enough space on which such a function could be printed.

5.1   The Variation Method

The SE is obtained from the Schrödinger Condition using the variational approach which was sketched in Chapter 2 and is the most complete derivation in the sense that it allows all possible values of the state function (Ψ) consistent with the boundary conditions (also generated by the same variational calculation). Any constraint placed upon the allowed variations in Ψ would, obviously, generate some analogue of the SE which would, when solved, yield something approximating to the solution of the SE — a state function Ψ' and an energy E', say. There is an obvious limitation on these solutions in the sense that the energy E' must always be higher than the energy E:

$$E' > E$$

since the constrained function can never be lower in energy than the fully optimised one.

Thus, the way is open for a foolproof method of obtaining approximate solutions to any SE. Simply place constraints on the action of the variation principle until the resulting equation can be solved and minimise the resulting functional by optimising the values of any parameters contained in the constraints. That the energy of the approximate function must always be higher than the experimental value ensures that one is homing in on the best approximation. The remaining question is how to choose the constraints so that the resulting approximation gives sensible results. Clearly, if the energy E is known from the experiment, the problem is made much easier, vary the constraints so that E' is as close as possible to E. The choice of constraints depends on having some intuitions about what is happening in the system under study, there are two main classes of method which are not completely independent:

- What one might call 'model constraints' — intuitions about the structure of the system under study.
- 'Numerical approximations' — choice of approximating functions and the parameters which such functions contain to be optimised.

This method dominates the uses of Schrödinger's Mechanics (SM) in quantum chemistry: the Hartree–Fock methods and its extensions. The physical interpretation of the resulting functions and their energies is straightforward since these quantities are known, from much experience, to be more or less accurate approximations[1] to the unknown exact solution. Needless to say in any application of the variation method the main guide to the best choice of functions is experience; systematic trial and error.

It might be useful here to list the main approximations used in the Hartree–Fock method:

- Model constraints:
  Any state function of a many-electron system must be anti-symmetric with respect to the exchange of any two electrons — electrons are fermions.

  Each electron in a many-electron system can have its own separate state function and therefore its own electron distribution. These two together mean that the overall state function is a single determinant of the one-electron state functions.

- Numerical constraints:
  The state functions of each electron can be approximated by a linear combination of functions which satisfy the boundary conditions necessary (exponentials vanishing at infinity).

  Both the linear expansion coefficients and any (nonlinear) parameters in the functions are optimised.

The result of applying these constraints is an equation — the Hartee–Fock equation — whose solutions are the state functions for each one of the electrons. This method can be extended by removing the second of the model constraints to generate more and more accurate approximations in the certain knowledge that the fewer constraints there are the more reliable will be the results.

The main weakness of this method is that it is difficult to get accurate approximations to small energy differences which are often very small and, when calculated as the difference between two large numbers, are often in the rounding error of the large numbers. Clearly what is needed is a method which calculates energy differences directly.

5.2   Perturbation Theory

The aim of this approach is to get a solution to an intractable SE by starting from a SE which *can* be solved and is, in some sense, 'close' to the insoluble SE; the difference between the two is a 'perturbation' on the soluble SE. The clue is in the name, the method is at its best when the difference between the effect of the perturbation is a small modification to the solutions of the basic soluble SE; for example, the change in energy due to the perturbation might be small compared to the energy of the unperturbed system.

The starting point is a SE

$$\hat{H}_0 \psi_0^{(i)} = E_0^{(i)} \psi_0^{(i)} \tag{5.1}$$

for which all the solutions $\psi_0^{(i)}$ and $E_0^{(i)}$ are known and an intractable SE

$$\hat{H}' \psi' = E' \psi' \tag{5.2}$$

for which a solution is sought where, to avoid too many subscripts and superscripts, ψ' is the lowest state of the perturbed system. The technique is to write the operator $\hat{H}'$ as:

$$\hat{H}' = \hat{H}_0 + \lambda \hat{V}$$

The perturbation $\hat{V}$ has been give the honour of wearing a hat but is often a simple potential. λ is a parameter which ideally has a physical interpretation, for example a field strength.

The assumptions (axioms) involved in the method are:

- The energy and state function may both be expanded as power series in the parameter λ

$$E' = E_0 + \lambda E_1 + \lambda^2 E_2 + \lambda^3 E_3 + \cdots \qquad (5.3)$$

$$\psi' = \psi_0 + \lambda \psi_1 + \lambda^2 \psi_2 + \lambda^3 \psi_3 + \cdots \qquad (5.4)$$

Inserting these two expressions into the perturbed Equation (5.2) obviously gives an equation very much more complex than the original:

$$(\hat{H}' + \lambda \hat{V})(\psi_0 + \lambda \psi_1 + \lambda^2 \psi_2 + \lambda^3 \psi_3 + \cdots)$$

$$= E'(\psi_0 + \lambda \psi_1 + \lambda^2 \psi_2 + \lambda^3 \psi_3 + \cdots) \qquad (5.5)$$

- It is assumed, and this is the key axiom of the perturbation method, that this equation is an *identity* in λ, i.e. *not* an equation to be solved for the ψ_i. With this axiom one may generate a whole series of linked *equations* for the ψ_i by equating the coefficients of like powers of λ. How are these equations to be solved?
- Obviously, as equations, they can be solved by any method whatsoever, but in view of previous considerations the most hopeful method would be to expand each ψ_i as a linear combination of some known functions. It must be stressed that the functions used in this expansion can be any (ideally) complete set and its choice is a combination of computational convenience and possible physical interpretation.
- Also, the number of equations for which the solutions are needed depends on the magnitude of the $\lambda \hat{H}'$ compared to the energy

$E_j^{(0)}$. It is also assumed without proof that the set of solutions of the power-series equations *exist and converge*.

The way to proceed, in view of the fact that algebra is easier than analysis, is to expand any or all of the hierarchy of the functions ψ_j as a linear expansion of a set of functions which span the whole space and have the same boundary conditions as $\psi_j^{(0)}$. There are several factors involved in choosing an optimum set of expansion functions since, despite the perturbation technique, in common with the variation method, it is necessary to be able to calculate integrals involving the expansion functions and the energy operators $\hat{H}$ and $\hat{V}$ which are involved in the algebraic operations. The criteria for the choice of functions are, therefore, often more mathematical than physical.

It is particularly mathematically convenient if these expansion functions are orthogonal among themselves *and* orthogonal to the unperturbed function ψ_j^0; this simplifies the resulting algebra enormously, one might almost say that, without this orthogonality, perturbation theory is not feasible as a practical technique. The most mathematically obvious set of such functions are the ψ_k^0 for $k \neq j$ since, as the functions which solve the original unperturbed equation form a complete orthogonal set themselves, they automatically satisfy the orthogonality requirements.

This mathematically straightforward approach, which, mathematically, is simply the most convenient one, has given rise to a most unfortunate tendency to give the terms in the energy expansion corresponding to this particular expansion of the perturbed state function ψ' (the $E^{(n)}$) *physical interpretations* which are wholly spurious. Perhaps the most unfortunate example of this false physical interpretation is the idea of 'virtual particles'; ghostly particles gliding about in space around 'real' particles. These alleged particles arise from terms in the perturbation expansion which, among other undesirable properties, do not obey the conservation of energy. Rather than realising that this is the result of giving a physical interpretation to the terms in the expansion, theorists have retained the interpretation and cast about for a reason for the unfortunate breaking of a fundamental law. The virtual particles have extremely short lifetimes so that there is absolutely no possibility of measuring their energies or, indeed, any of their properties. So, using a rather disputed relationship, the so-called fourth uncertainty principle,[a] which would use the non-measurability of the departures from the conservation of energy; using that well known principle: an unsavoury happening only becomes a scandal when someone knows about it.

There is no *physical* requirement that one use the ψ_k^0 as expansion functions, in fact, there is no theoretical requirement that the expansion technique must be used at all; one possibility would be to solve the perturbation equations variationally or even, if one had the time and patience, numerically. More realistically, the functions used in a linear expansion to solve these equations could be any convenient set — they do not have to be an orthonormal set or, indeed, orthogonal to $\psi_j^{(0)}$. If the terms arising in the expansion are given a physical interpretation, then there will be different 'phenomena' associated with each expansion set which is clearly ridiculous.[2] Equally obviously, there is no theoretical way of partitioning the Hamiltonian operator in any fixed way. Equally important, there is no 'guidance' theorem analogous to the

$$E' > E$$

of the variation method; no guarantee that the result gets better (closer to experiment) as the length of the expansions is increased.

In atomic physics, for example, if one were to use the unperturbed (hydrogen-like) functions of the atom without electron interaction the result would not be worth the huge effort involved; instead, one gets a good Hartree–Fock state function, perturbs it with the known deficiencies of the HF method and, finally, perturbs this corrected function with operators representing the properties of interest; a physical, rather than a mathematical priority. To take the simplest possible example, if one wants to find an approximation to the ground-state of a hydrogen atom in a constant electric field, the 'obvious' function to use for $\psi_0^{(0)}$ is the ground-state $(1s)$ function of

---

[a]See Section 5.3.

the unperturbed atom. The corrections which must be made to this function to allow for the presence of this field are intuitively obvious; the spherically-symmetric electron distribution will be polarised — skewed to one side or the other. Using the excited-state functions to expand the perturbed state does not work at all well for the obvious reason that excited-state functions have more and more diffuse electron distributions as their energy increases and these functions will not change the perturbed distribution at all well; the required corrections must be made in the same general region of space as the one associated with $\psi_0^{(0)}$. If one uses a single function of the next-higher energy ('2p') and variationally optimises the *'size'* of both the ground-state and the $2p$ functions a much clearer picture of what has happened can be obtained. In this simple case both the unperturbed and the perturbed states are time independent; there are no 'processes' going on; there has not been nor is there any excitation from the $1s$ to the $2p$ state of the unperturbed H atom.

There is always the possibility that the required perturbed function cannot be expanded by the polynomial method of perturbation theory; that is the failure of the most basic assumption. This is the case for the expansion of the ground state of the H_2^+ ion by using $\lambda \hat{V} = 1/R$ (R is the internuclear distance) when the expansion is the set of eigenfunctions of the H atom; there is a logarithmic term in the exact expansion. More confusingly, there are cases, even when the perturbation series seems to converge, for which it can be proved analytically that it does not and, although the numbers seem to be right, if they are given a physical interpretation, it is bound to be doubtful or erroneous. The fact is that the original perturbation method was 'designed' to be used when the perturbation is small compared to the unperturbed energy; in Elliott. H. Lieb's words:

> We are looking for small effects, called 'radiative corrections', and these effects are like a flea on an elephant. Perturbation theory treats the elephant as a perturbation of the flea.
>
> E.H. Lieb, *Physica A*, **263**, (1999), p. 491.

In cases where the perturbation effects are small, the theory is an excellent tool and works well in situations where the variational method is, in practice, useless. Performing a variational calculation on both the unperturbed and perturbed systems and using the difference has been compared to weighing the captain of a cruise liner without him aboard and after he has boarded and subtracting the results; the required number is way down in the rounding error of any possible measurements.

All this is perfectly well understood in theoretical chemistry (and electron theories in general) precisely because there is a body of chemical knowledge built up over at least a century and the intuitions about the structure and transformations involved are constantly being improved; hence the existence of many shelf-miles of chemical literature. It is now possible to be able to calculate the paths of processes and electron distribution changes which are presently beyond experimental detection with a good deal of confidence. But in modern field/particle theories this experience and these intuitions are absent for perfectly valid reasons; the mechanisms of many of the processes are simple not adequately known. In these cases the interpretation of the terms in the perturbation expansion has, unfortunately, had to do duty as a physical interpretation but with no evidence that these interpretations are even meaningful, let alone correct.

5.3 The Fourth Uncertainty Principle

There comes a time when, in any work on SM, one must address the question of the so-called 'fourth uncertainty principle' mentioned in the justification of 'virtual particles' in the previous section; this invites the question: is there an uncertainty 'principle' involving energy and time analogous to the uncertainty theorems relating the distributions of a coordinate and its conjugate momentum mentioned in Section 2.6? That is, does the relationship often written

$$\Delta E \Delta t \geq \frac{\hbar}{2} \tag{5.6}$$

presumably mean

$$\sigma_E \sigma_t \geq \frac{\hbar}{2} \tag{5.7}$$

where the ΔE and Δt are not clearly defined but the σs are the standard deviations of the distributions of E and t. Do these inequalities actually exist? And, if they do exist what do Equations (5.6) and (5.7) mean?

It has been stressed many times that the referent of SM, which is a *theory*, is systems of particles not measurements made on such systems so that Equation (5.6) cannot be interpreted as a relation between measurements of differences of energy and of time even though this was how the original 'principle' was justified by the doubtful method of 'thought experiments' and a few actual experiments. It is necessary to know whether or not there are *operators* $\hat{E}$ and $\hat{t}$ related to energy and time which play a similar role to the operators $\hat{p}_j$ and the conjugate $\hat{q}^j$ as

$$\hat{p}_j = -i\hbar \frac{\partial}{\partial_j}$$

$$\hat{q}^j = q_j \times$$

It is well-known that, $\hat{E}$ the energy operator[3] does exist[b] and appears in the SE as

$$\hat{E} = i\hbar \frac{\partial}{\partial t}$$

So the theory hinges on the existence of an operator $\hat{t}$ analogous to the $\hat{q}^j$ which is simply scalar multiplication[c] by t. This may look like a sophistry since multiplication by t (or, of course, by any q^j) is always possible but in SM it is not so simple since only particle variables can be associated with an operator. The mathematical difference between an operator $\hat{t}$ and simple scalar multiplication by t is obviously non-existent; the difference is one of *interpretation*; is t a particle variable? And, incidentally, what might be the *mean* of the operator $\hat{t}$ which σ_t is the standard deviation from?

---

[b] And it is not the operator $\hat{H}$!
[c] We must go back to primary school to remember the explicit use of $\times$ for multiplication!

It might be useful to look again at the two possible formulations of CPM each of which have been claimed to be the precursor of SM and see the particle variables involved in each:

(1) There are two related formulations of CPM to be considered here, Lagrange's equations, and Hamilton's canonical pairs.

   The solutions of both of these systems are similar, in particular both systems generate the *trajectories* of particles, the $q^j(t)$. It is clear that the q^j are particle variables while t is not; it is a parameter on which each q^j depends. With this interpretation, in SM $t\times$ is definitely not an operator; a dynamical variable of the particles.

(2) The other method is via the HJE.

   In this case the situation is quite different. The solutions of this equation form the set of all possible trajectories for the particles, the q^j are not functions of t, in fact the q^j and t are a set of independent coordinates which span the spacetime continuum in which the trajectories exist.[d] Of course, when one has the solutions of the HJE and suitable initial data one can calculate the trajectories of the particles which may still be written $q^j(t)$ but these q^js are not the same as the original (q^j, t) system of coordinates. This is an extremely unfortunate confusion of notation which is endemic in the literature. Ideally, when the coordinate system (q^j, t) is used in the HJE there should be a different notation used for the actual trajectories when computed — $y^j(t)$, say — to avoid the duplication of interpretation. When the HJE is solved all the points in the space spanned by the q^j are on one or more of the trajectories $y^j(t)$.

In both cases it cannot be said that t is a particle variable, in the Hamilton–Jacobi Equation (HJE) itself neither the q^j or t are properties of the particles, they define points in spacetime. When particular particle trajectories y^j are actually computed, then the set of all $y^j(t)$ is the same as the set of all the $q^j(t)$ and, again, t is a parameter, not a particle variable. In the relativistic case if

---

[d]A point which was emphasised in Chapter 2.

the particles are in relative motion, each one apparently has, from a given reference frame, its own time so, apparently t for each particle is needed to specify it fully. A little thought shows that, if there are particles in the same reference frame and, in fact, in general particles do not each have their own unique time. In any case, if the particles interact then there is always the problem, as yet unsolved in SR of how to deal with particles of constantly changing velocity.

There is a simpler way to look at the problem; in the case of the $q^j(t)$ there is a mean value of the distribution of that q^j which is easily calculated by

$$\langle q \rangle = \int_a^b \psi^* \hat{q} \psi dV = \int_a^b \psi^* q \times \psi dV$$

where a and b are the limits of the distribution of ψ in the volume V. It is this mean from which the standard deviations of σ_q and a similar σ_p are computed. It is obvious that there is no mean value of *time*; what could it possibly be; half past three, quarter to ten?

The conclusion is obvious, there is no fourth uncertainty 'principle' as given by Equation 5.7. Equation 5.6 has found some experimental use however, but this involves giving the quantity Δt the value of a half-life not the standard deviation formed from a state function. There has to be a limit to the use of doubtful mathematical formula to 'explain things away' rather than to explain them and surely the arbitrary use of the 'fourth uncertainty principle' is that limit.

Endnotes

[1] There are pitfalls in the variational optimisation method; how to ensure that the minimum found is a global minimum not a local one.

[2] If one were to expand the interaction of two molecules using Tchebechev polynomials, for example, it would be absurd to regard the coefficients in this expansion as indicating the phenomena involved.

[3] There are considerable disputes about whether the Hamiltonian operator $(\hat{H})$ or the operator $i\hbar\partial/\partial t$ is the correct energy operator. One persistent claim that it cannot be the energy operator since its eigenvalues are not quantised; they form a continuum whereas it is fundamental to SM that the energy *is* quantised. This has been discussed in Section 4.4; in fact, of course the momentum operators are of exactly the same form as the energy operator and have the same continuum of eigenvalues which is never challenged. It must be stressed again that SM does not apply to all distributions and energies of particles but only to those whose motion is *constrained* in some way. Constraints are usually the presence of a potential field or the imposition, in the case of momentum operators, of limits to the motion of the particle. The latter is also, in fact, the presence of a potential field but is not presented as such due to the mathematical problems associated with impacts and fields of infinite or no gradient. In the case of the energy operator $i\hbar\partial/\partial t$, the constraint is provided by the spatial terms in the Hamiltonian.

Chapter 6

Spin in Schrödinger's Theory

The first introduction of multi-component state functions was by Pauli who, familiar with the representation of electron spin operators as 2×2 matrices, incorporated these matrices as a 3D vector into the SE in the presence of an electromagnetic field. This was done empirically and it generated a Hamiltonian operator which had a correct term representing the dipole/magnetic field interaction of the 'spinning electron'. This chapter shows that this interaction follows naturally from a more explicit description of ordinary 3D space and does not have to be an empirical addition to SM.

6.1 The Square of Momentum

In this section, notwithstanding the warnings on page 28, only Cartesian coordinates will be used to avoid a confusing clutter of notation which is not relevant to the main thrust of the argument. Thus, the conventional way to form the Hamiltonian in terms of Cartesian coordinates (x^i, say, for a single particle, mass m) and momenta p_i is to retain the spatial coordinates unchanged and use the substitution:

$$p_i \rightarrow -i\hbar \frac{\partial}{\partial x^i} = \hat{p}_i \quad \text{(say)}$$

Then the 'kinetic energy operator' for each x^i is given by:

$$\frac{1}{2m}\left(-i\hbar \frac{\partial}{\partial x^i}\right)^2 = -\frac{\hbar^2}{2m}\frac{\partial^2}{\partial x^{i2}} = -\frac{\hbar^2}{2m}\hat{p}_i^2$$

leading to the familiar shorthand for this operator as

$$-\frac{\hbar^2}{2m}\nabla^2$$

Notice here that it is the *square* of the three components which generates the correct operator and *not* the square of the modulus of those components so that the correct *sign* for the operator is obtained.[1]

Now, if this argument is applied to the full expression for the momentum, that is the momentum *vector* is squared, not just the sum of the squares of its components, we get a very different result. While

$$\sum_{i=1}^{3}(\hat{p}_i)^2 = |\hat{\boldsymbol{p}}_i|^2$$

is the conventional result, the square of the *vector* $\hat{\boldsymbol{p}}$ gives:

$$\hat{\boldsymbol{p}}^2 = \sum_{i,j=1}^{3} \hat{p}_i\hat{p}_j\boldsymbol{e}_i\boldsymbol{e}_j$$

containing nine terms involving the bilinear expressions in the three (Cartesian) unit vectors $\boldsymbol{e}_i$ and, so long as these unit vectors are left as such, the standard methods of vector algebra leads to no further progress.

But there is a more powerful and physically transparent method to represent the properties of space which enables the squares of vectors to be obtained and given a physical interpretation: what is now known as Geometric Algebra (GA) has been discovered and rediscovered several times over the last two centuries as quaternions, Grassman algebra and Clifford Algebra and is now receiving increasing attention and application in many branches of physics thanks to the untiring work of David Hestenes. For our purposes, working in ordinary 3D Euclidean space, we only need one of its basic formulas for the general product of two vectors:

$$\boldsymbol{ab} = \boldsymbol{a}\cdot\boldsymbol{b} + \boldsymbol{a}\wedge\boldsymbol{b}$$

which can be translated into 'normal' vector notation[2] as

$$ab = a \cdot b + ia \times b$$

where:

$i = \sqrt{-1}$ in vector algebra and the dual operator of GA,[a]

$a \cdot b$ is the usual scalar product in both GA and standard vector usage, and

$a \wedge b$ is the GA "outer product" which is — in 3D — related to the vector product $a \times b$ by the above rule.

It is easy to see that, for ordinary vectors:

$$p \wedge p = ip \times p = 0$$

so that

$$pp = p \cdot p = |p|^2$$

and the same result holds for all vectors *provided that the individual components of the vectors involved in the product commute*; for example:

$$\nabla\nabla = \nabla^2 \quad \text{because} \quad \frac{\partial}{\partial x^i}\frac{\partial}{\partial x^j} = \frac{\partial}{\partial x^j}\frac{\partial}{\partial x^i}$$

which, of course, means that the GA expression for the square of the classical momentum operator and the SM kinetic energy operator are the same as they are in ordinary vector notation; apparently, nothing is gained from the additional generality of GA; but nothing is lost, the method works.

The interesting case occurs when a vector potential (A, say) is present and the kinetic energy operator is replaced by:

$$\frac{1}{2m}\left(p - \frac{e}{c}A\right)^2 \rightarrow \frac{1}{2m}\left(-i\hbar\nabla - \frac{e}{c}A\right)^2 \tag{6.1}$$

---

[a]It generates the vector normal to the plane $a \wedge b$ defined by the two (non-colinear) vectors a and b.

where standard (S.I.) units are used in order to display the forms of the resulting expressions in a more familiar way. Although this expression is the square of a vector it is not the same as the standard scalar product form since, for example, $\nabla \cdot \boldsymbol{A} \neq \boldsymbol{A} \cdot \nabla$ and $\nabla \times \boldsymbol{A} \neq -\boldsymbol{A} \times \nabla$.

This expression is easily expanded using the above rules:

$$\boldsymbol{ab} = \boldsymbol{a} \cdot \boldsymbol{b} + \boldsymbol{a} \wedge \boldsymbol{b}$$

$$\boldsymbol{a} \wedge \boldsymbol{b} = i\boldsymbol{a} \times \boldsymbol{b}$$

and noting that:

$$\nabla \times \boldsymbol{A} = \boldsymbol{H} \quad \text{(The magnetic field vector)}$$

$$\nabla \times \nabla = 0$$

$$\boldsymbol{A} \times \boldsymbol{A} = 0$$

$$\text{and} \quad \nabla \cdot \boldsymbol{A} = 0 \quad \text{(Coulomb gauge)}$$

The result is:

$$\frac{1}{2m} \left\{ -\hbar^2 \nabla^2 + \frac{e^2}{c^2} \boldsymbol{A} \cdot \boldsymbol{A} - \frac{\hbar e}{c} \nabla \times \boldsymbol{A} - \frac{\hbar e}{c} (\boldsymbol{A} \cdot \nabla + \boldsymbol{A} \times \nabla) \right\}$$

That is, grouping together the terms involved in the conventional scalar product to emphasise the difference:

$$\frac{1}{2m} \left(-i\hbar\nabla - \frac{e}{c}\boldsymbol{A} \right) \cdot \left(-i\hbar\nabla - \frac{e}{c}\boldsymbol{A} \right) - \frac{\hbar e}{2mc} \boldsymbol{H} - \frac{\hbar e}{2mc} \boldsymbol{A} \times \nabla$$

we see that there are two additional terms when compared to the usual result obtained by simply summing the squares of the components; each multiplied by the Bohr magneton:

$$\mu_B = \frac{\hbar e}{2mc}$$

one linear in the magnetic field vector $\boldsymbol{H}$ multiplied by the Bohr magneton and a less familiar one. This term, involving the same Bohr magneton and $\boldsymbol{A} \times \nabla$ is not so simple to interpret. The simplest case is when A generates a constant magnetic field, then:

$$A = \frac{1}{2} H \times r$$

which makes this term into:

$$\frac{\mu_B}{2} H \times (r \times \nabla) = \frac{\mu_B}{2} H \times L$$

where $L = r \times \nabla$ is the angular momentum operator.

6.2   Schrödinger or Pauli?

Having obtained a Hamiltonian operator using the methods of GA and translated the result into traditional vector language, the question arises:

> What are the functions which either solve the associated SE or are suitable to use in seeking an approximate solution?

The problem is that parts of this Hamiltonian operator are scalars, (∇^2 the scalar product terms in ∇ and A and any electrostatic potential) and parts (the new ones involving the magnetic field) are vectors. There is no such thing as an 'identity vector' which might be used to multiply the scalar terms to save the situation so we are at an *impasse*. We must turn to GA again.

The above expression for the Hamiltonian is expressed in normal vector notation for familiarity, but in order to interpret these terms we need to obtain a concrete algebraic representation of the GA vectors by means of an explicit representation of the three unit vectors, something that is not possible in ordinary vector algebra. The unit vectors e_i are basic units defined only by their rules of combination:

$$e_i \cdot e_j = \delta_{ij} \quad \text{and} \quad e_i \times e_j = \epsilon_{ijk} e_k$$

(their interpretation as orthonormal unit vectors). The above Hamiltonian has been obtained using the abstract rules of GA which are not obeyed by the standard e_i so that we need a concrete representation which does.

There *is* an explicit mathematical representation of the GA equivalent of the e_i. The three orthonormal vectors are represented by the three matrices reintroduced into quantum theory in the last century by Wolfgang Pauli. These matrices are just that — matrices — and obey all the familiar rules of matrix manipulation. Thus, the e_j have no properties other than those explicitly assigned to them , unlike the σ_j which are mathematical objects of a familiar kind. In particular, there *is* an identity matrix which occurs naturally in the development and the scalar and vector parts are put on a common footing. These representations are distinguished from the primitive unit vectors e_i by their notation[b] σ_i where:

$$\sigma_1 = \begin{pmatrix} 0 & 1 \\ 1 & 0 \end{pmatrix} \quad \sigma_2 = \begin{pmatrix} 0 & -i \\ i & 0 \end{pmatrix} \quad \sigma_3 = \begin{pmatrix} 1 & 0 \\ 0 & -1 \end{pmatrix}$$

If the e_i are replaced by the σ_i and due regard is given to the order of the matrix products, the expansion of the quantity p^2 can be computed explicitly to give:

$$p^2 = (p_1^2 + p_2^2 + p_3^2) \begin{pmatrix} 1 & 0 \\ 0 & 1 \end{pmatrix}$$

by noting that:

$$\sigma_i^2 = \begin{pmatrix} 1 & 0 \\ 0 & 1 \end{pmatrix}$$

and

$$\sigma_i \sigma_j + \sigma_j \sigma_i = \begin{pmatrix} 0 & 0 \\ 0 & 0 \end{pmatrix} \quad i \neq j$$

This method[c] yields the same result as $\hat{p} \cdot \hat{p}$ with the important addition that *the operator is now defined as the product of the usual 'kinetic energy' multiplied by a two-component matrix.*

---

[b]No vector symbol, since they are manipulated by the normal methods of matrix algebra.

[c]Naturally, this representation of the σ_i is not unique, any unitary transformation

$$\sigma_i' = U^\dagger \sigma_i U$$

gives a "rotated" set of basis vectors with the same properties.

What this means is that, when due regard is given to the rule for generating the Hamiltonian operator by squaring the *vector* momentum, not simply its components, the SE is, in fact, what one might call a "degenerate" two-component equation which has two copies of the same spatial equation, one for each matrix dimension.

Obviously, the addition of an electrostatic potential energy term to the kinetic energy term does not affect the general result.

Now, the same procedure is carried through but with the inclusion of a vector potential as above

$$\left(-i\hbar\frac{\partial}{\partial x^i} - \frac{e}{c}A_i\right)$$

where

$$\nabla = \sigma_1\frac{\partial}{\partial x_1} + \sigma_2\frac{\partial}{\partial x_2} + \sigma_3\frac{\partial}{\partial x_3}$$

$$\boldsymbol{A} = A_1\sigma_1 + A_2\sigma_2 + A_3\sigma_3$$

The term explicitly involving the magnetic field, when written out in full is

$$\mu_B(H_1\sigma_1 + H_2\sigma_2 + H_3\sigma_3)$$

where

$$\mu_B = \frac{e\hbar}{2mc} \quad \text{the Bohr Magneton}$$

This is the term which is traditionally introduced empirically and written

$$\mu_B\boldsymbol{H}\cdot\boldsymbol{\sigma}$$

wrongly using the σ_i as *components* of a vector rather than their true interpretation as the orthonormal basis vectors for Cartesian space. This is clearly what has been called "spin-field" interaction. However, no mention of spin has been made in this derivation and it is not necessary to give the particle (which could have any charge, not necessarily the electronic charge) a magnetic moment. The Bohr magneton appears naturally in what seems to simply be part of the interaction of a charged particle with a vector field.

Returning to the second of the new terms, but this time, as above, with the constraint of $\boldsymbol{A}$ representing a constant magnetic field, it may be written out in full temporarily using the traditional determinantal form with (x, y, z) notation:

$$\boldsymbol{H} \times \boldsymbol{L} = \begin{vmatrix} H_x & H_y & H_z \\ L_x & L_y & L_z \\ \sigma_x & \sigma_y & \sigma_z \end{vmatrix}$$

It can be seen to involve the magnetic field $\boldsymbol{H}$, the angular momentum $\boldsymbol{L}$ and (hidden in the vector notation) the "spin" operators σ_i. In the historic terminology, it looks like magnetic-field-induced spin-orbit coupling.

There is, however, one further complication; in using the expression

$$ab = a \cdot b + a \wedge b = a \cdot b + ia \times b$$

no account has been taken of the GA nature of the i which now occurs in both the equation above and in the 2×2 matrix representation of the unit vectors. The "number" i also occurs as the "matrix" i

$$i = \begin{pmatrix} i & 0 \\ 0 & i \end{pmatrix}$$

which, in GA, is the operator which turns GA quantities into their "duals". The case in point here turns the unit vectors σ_j into unit planes in 3D. So, to be on the safe side, it probably is better to replace the occurrences of the "number" i into the 2×2 matrices thus making the entities which result from the "vector product" into *planes* in 3D. This enables a more smooth transition into the 4D spacetime where the vector product is not defined. The only question which remains is the nature of the i which is involved in the energy operator $i\hbar\frac{\partial}{\partial t}$. Clearly, multiplying the matrices by either the number i or the matrix i makes no difference to applications of the non-relativistic spin SE, but it would be better to have a clear idea of the interpretation of the important energy operator.

6.3   Comparisons

The real question about the "inclusion of spin" into quantum theory lies not so much in forming an equation which has more than one component — whether two or four — but what evidence that equation provides of what is called the spin or magnetic moment of the particle without introducing the existence and size of this effect arbitrarily.

It has been found that, entirely within the *ansatz* of the original formulation of Schrödinger's mechanics (SM), the "spin-field" term can be shown to be qualitatively and quantitatively correct. The two changes which must be made to Schrödinger's original paper of 1926 are:

(1) The result of the variational derivation that the kinetic-energy-related term in the quantum Hamiltonian operator contains ∇^2 and this is obtained by squaring and summing the Cartesian momentum operator components must be replaced by squaring the momentum *vector*.

(2) The unit Cartesian vectors must be given an explicit (geometric algebra) representation rather than the conventional symbolic representation.

When this is done, it indicates and effectively proves several things:

- The dipole-field term in the SE is entirely a "quantum effect". The fact that the square of the classical momentum vector — including the vector potential — is identical to its squared modulus (scalar product) means that the terms arising from the "wedge product" are identically zero.
- No additional *ad hoc* assumptions about the charged particles' "intrinsic" motion or magnetism need be made.
- The dipole-field interaction arises when a more careful description of the properties of space are used rather than some electromagnetic property of a charged particle other than its charge.
- This derivation only requires the particle to have an electric charge (q, say); the sign or magnitude of the charge is not specified, the particle need not be an electron; the apparent dipole is always of

the same form, i.e.

$$\frac{\hbar q}{2mc}$$

corresponding to 'spin 1/2'.

- The existence of an energy of the form of a dipole-field interaction is not a relativistic effect. In fact, it is extremely useful to be able to separate the familiar, genuine, relativistic phenomena — like variation of mass with velocity — from 'spin' which has no relativistic heritage.
- And, finally, it gives satisfaction to remove an empirical addition from a familiar form of the SE.

Endnotes

[1] The negative sign arises in the full variational derivation — where the square of the modulus of the momentum, the correct term for the kinetic energy — due to an integration by parts.

[2] This relationship is known in QM circles as the Pauli identity.

Chapter 7

Relativistic Equations

There are quite a number of obstacles to be overcome before any attempt at a transition to a relativistic equivalent of the Schrödinger Equation (SE) can be attempted. The Special Theory of Relativity (SR) contains no *dynamical* laws, what it does is place restrictions — inherited from electromagnetic theory — on the existing laws of CPM, i.e. Newton's laws.[1] That is, SR is basically about Kinematics (properties of space) rather than mechanics.

The areas about which the theories of SR and Schrödinger's Mechanics (SM) were developed could hardly be further apart. SR has its main use in 'large-scale' physics and its effects are at a maximum in treating macroscopic systems at very high constant relative velocities while SM is effective at the microscopic scale with variable velocities due to the presence of constraints on motion. In a particle-mechanics context, SR is all about particles and associated frames of reference moving at constant relative velocities. Thus, it is hardly to be expected that it will mesh cleanly with SM since the most obvious area of applicability of SR is that of the motion of free particles in the absence of any field of potential. Such motions are, as we have seen, not quantised and of minimal physical interest although of some pedagogical value when artificial boundary conditions are applied. The uses of the SE are overwhelmingly concerned with particles whose motions are constrained by potential fields. The absence of any potential means, naturally, that there is no room here for Newton's all-important third law of particle mechanics ('$\boldsymbol{F} = \dot{\boldsymbol{p}}$')

and only the first law ('in the absence of a force, a particle moves at constant velocity') need be invoked. However, there is great demand to see that all of mathematically articulated physical science is consistent and so it is worth seeing how far one can get in achieving this aim.

The most daunting of the difficulties associated with the attempt to follow the path taken in earlier chapters in forming a quantum-mechanical SR-compatible analogue of SM is the breakdown of the whole

$$\text{Newton} \to \text{Lagrange} \to \text{Hamilton–Jacobi} \to \text{Schrödinger}$$

approach. The reason for this breakdown is quite clear; in the variational "homogeneous case" where the set of coordinates is $(x^j) = (q^j; t)$ $j = 1, n)$ say, such that the x^j are all on the same footing, there are, in the Lagrangian, terms involving $dx^j/dx^{n+1} = dq^j/dt$ while in the classical case where t is the parameter on which the q^j depend, there are none of the type dq^j/dq^ℓ. That is, while the relativistic 'coordinates' are *linearly* independent, they are not independent nor are they in any way physically all equivalent. Time, whether linked to a given reference frame or proper, always has a different significance from spatial coordinates. As we shall see, this problem arises later in another guise in the energy-momentum vector and throughout all attempts to make the standard analytical methods of CPM work satisfactorily in a relativistic environment.[2] Naturally, it is useful to see the severity of these difficulties on CPM and, by implication, on SM.

It is also important to see to what extent the apparently insurmountable obstacles to a relativistic quantum mechanics (QM) can be overcome by taking seriously the Minkowski metaphor of 4D spacetime and extending the 3D mathematics to four dimensions.

7.1   Hamiltonian and Lagrangian Functions

When faced with a new situation, it is usual in quantum circles to approach the problem of the correct choice of Hamiltonian operator by guessing a likely expression for the classical energy of the system in terms of some convenient coordinates and replace each momentum component p_i by $-i\hbar(\partial/\partial q^j)$ with due regard to what was said on page 28 about the initial use of Cartesians. But the route to the Hamiltonian *function* of CPM is *via* the Lagrangian function $L(q^i, \dot{q}^j; t)$ and the relations:

$$p_j = \frac{\partial L}{\partial \dot{q}^j}$$

$$H = p_j \dot{q}^j - L$$

These assumptions do not normally cause any difficulties in non-relativistic mechanics but when the constraints of SR are applied the situation is more complex.

However, in this Ptolomaic spirit of looking for a way to get some kind of method which gives reasonable answers, we can start in the middle of the method of Chapter 2 by looking for an expression for the classical SR 'Lagrangian' which *does* give the all-important definition:

$$p_j = \frac{\partial L}{\partial x^j}$$

for the standard ('spatial') momentum components.[3]

In standard CPM, the expression for the kinetic energy is obtained by assuming that the work done on a particle by a field of force ($\boldsymbol{F}$, say) over a path described by the vector $\boldsymbol{r}$ appears as the kinetic energy T

$$T = \int_{\boldsymbol{r}_a}^{\boldsymbol{r}_b} \boldsymbol{F} \cdot d\boldsymbol{r}$$

and, when Newton's law,

$$F = m_0 \frac{d^2 \boldsymbol{r}}{dt^2}$$

is used, we have the familiar expression

$$T = \frac{1}{2} m_0 \left(\frac{d\boldsymbol{r}}{dt} \right)^2$$

in which the mass of the particle (m_0) is assumed to be constant. SR replaces the constancy of mass by the rule:

$$m = m_0 \left[1 - \left(\frac{v}{c} \right)^2 \right]^{-\frac{1}{2}}$$

and the — somewhat more involved — expression for the kinetic energy obtained by the above sequence becomes

$$T = m_0 c^2 \left\{ \left[1 - \left(\frac{v}{c} \right)^2 \right]^{-\frac{1}{2}} - 1 \right\}$$

Now, the Lagrangian for a particle in a time-independent field of force has potential energy $V(x^j)$, (say) and so the Lagrangian is

$$L = T - V$$

and

$$\frac{\partial L}{\partial \dot{x}^k} = \frac{\partial T}{\partial \dot{x}^k} \neq m\dot{x}^k = p_k$$

assuming, of course, Cartesian coordinates. Tolman[4] goes on to define a new form for the 'Lagrangian' which does enable the CPM relationship between L and p_j

$$L = -mc^2 \left(1 - \frac{\dot{x}^2}{c^2} \right)^{\frac{1}{2}}$$

which is the current, accepted, Lagrangian which does, in fact, generate the 'correct' intuitively acceptable form for the momentum

$$\frac{\partial L}{\partial \dot{x}}$$

which enables the CPM equations to be cast into a familiar form, but there are more fundamental difficulties which place more severe restrictions on the form of the CPM Lagrangian.

7.1.1 Covariant Lagrangian?

In CPM, the invariance of the *value* of the variational integral is

$$I = \int_a^b L(q^j, \dot{q}^j; t) dt$$

which, with respect to transformations of spatial coordinates, is clear to see, but SR imposes a more stringent requirement; the Lagrangian function itself must be of invariant scalar *form* with respect to transformations of coordinates and time. If, therefore, we make the very general assumption that the coordinates are not $(q^j; t)$ but the general set (x^j) and each one is a function of some parameter s (say), i.e. $x^j(s)$ using the usual notation where the parameter is s rather than t, then

$$\dot{x}^j = \frac{dx}{ds}$$

The putative Lagrangian will then be $L(x^j, \dot{x}^j)$ and the relevant integral required is

$$I = \int_{s_1}^{s_2} L(x^j, \dot{x}^j) ds \tag{7.1}$$

If we now assume that this integral is subject to a transformation of the parameter s to a new parameter $\sigma(s)$, say, then

$$\dot{\sigma} = \frac{d\sigma}{ds}$$

and the transformed integral becomes

$$I = \int_{\sigma_1}^{\sigma_2} L(x^j, dx^j \dot{\sigma}) \frac{d\sigma}{\dot{\sigma}}$$

and this is required to be equal to the original integral I which implies that L must be *linear* in the $\dot{x}^j$.

The most surprising result of this conclusion is that, if the momenta p_i of the system are obtained in the usual way by

$$p_j = \frac{\partial L}{\partial \dot{x}^j}$$

then the momenta are not proportional to the velocities and are, in fact, *independent* of the velocities and, in the SR Lagrangian are all *constants* (given by $\dot{x} = c$) where c is the velocity of light. Less obviously, the requirement that the Legendre transformation between L and H be reversible is no longer obtained, essentially because the Jacobian of the transformation is singular, involving second derivatives of the two coordinate systems which are zero.

Nevertheless, following this scheme we find that the 'Hamiltonian' under these conditions, that is

$$`H(x^j, p_j)' = p_j \dot{x}^j - L(x^j, \dot{x}^j) = 0 \tag{7.2}$$

and since, in fact, the L of Equation (7.1) is completely general and, specifically, has not been required to include any law of nature, Newton's, Schrödinger's or Einstein's, the above expression is a mathematical *identity* not an equation with a physical interpretation. To see that this is the case it is necessary to recall the confusion between the coordinates of space and those of a particle in space emphasised on page 13. The general Lagrangian as introduced in Equation (7.1) is a function of the coordinates of spacetime and the 'velocities' do not exist *as yet* since the variational problem of minimising L to generate the x^j as functions of σ and so $\dot{x}^j$ is merely formal. Only when the law of nature is supplied to this identity (the minimisation of the integral I in the equation) are the allowed trajectories defined as solutions of the implied Euler–Lagrange equations. Then and only then do the coordinates x^j become the coordinates of particles and the $\dot{x}^j$ become the velocities of those particles; in short, the x^j and the $\dot{x}^j$ become functions of σ (or *time*) on the allowed trajectories, not coordinates of spacetime.

But, when written out in full and the x^j are rewritten in pre-SR terms as (q^j, t), Equation (7.2) is 'somewhat' reminiscent of the CPM

Hamilton–Jacobi Equation (HJE)

$$``\sum_{j=1}^{3} cp_j = \hat{E}\,"$$

because the former right-hand-side of a non-relativistic SE (the energy) is already contained in the energy/momentum vector and may be restored to its traditional place in the SE. That is, it is of the form:

'left-hand-side is a function of the momenta and right-hand-side is the energy'.

Nevertheless, the Lorentz transformation is just a special case of this very general derivation and this means that in the relativistic case there seems to be no route to a meaningful 'Hamiltonian' and, *a fortiori*, no analogue of a HJE and therefore no transparent equivalent to Schrödinger's variational method.

Now one could regard this whole procedure as just a nuisance and try to *start* by defining some 'Hamiltonian' and work with this directly. But a covariance requirement on this Hamiltonian leads to the analogous conclusion that all the *velocities* must be independent of the momenta. Further, what were interpreted (wrongly, see Section 2.3.2) as the velocity components of the particle were independent of the momenta and were *constants*.

But, there may be a way around this inconvenient fact and by assuming that Newtonian mechanics and its development by Lagrange and Hamilton is just a *use* rather than the *source* of the Lagrangian method. In other words, investigate the *general* use of the variable t as another coordinate and that all these new coordinates depend on a monotonic, non-zero parameter σ,[5] say, as above. This is, of course, notwithstanding the fact that the Lagrangian may include dependence on terms like dq^j/dt but none of the form dq^j/dq^l. Basically, all that has been proved above is that the Legendre transformation does not work when time is treated as a coordinate; the insistence on covariance has sabotaged CPM before it is possible to use the HJE pathway to some analogue of the Schrödinger Equation.

In SM, since we require a Hamiltonian which must be expressed in terms of momenta and coordinates before trying this approach, it is useful to be sure what replaces the actual *momentum vector* in SR.

7.2   The SR Energy-Momentum 4-Vector

In SR the momentum 3-vector is replaced by the 4-vector of the three usual spatial components of momentum and a new component related to the time 'coordinate' (suitable scaled to obtain homogeneity of units — MLT^{-1}).

When the transition is made to SR — from 3D to 3D + time — then the 'energy momentum' vector and its quantum equivalent have the same properties as the 3D momentum operator; the eigenfunction 'equations' are exactly of the same type. In QM the energy operator is

$$i\hbar\frac{\partial}{\partial t}$$

and *not* the Hamiltonian operator. Indeed, in both classical and quantum mechanics it is the *equation* which links the energy and the Hamiltonian which inherits $F = ma$ by the above sequence. So that the relativistic energy-momentum operator is

$$P = (e_0 p_0, e_1 p_1, e_2 p_2, e_3 p_3) \quad \text{where } p_0 = \frac{E}{c}$$

which, when translated into operator form using the HJE method, gives

$$\hat{P} = \left(e_0 i\hbar\frac{\partial}{\partial ct} - e_1 i\hbar\frac{\partial}{\partial x^1} - e_2 i\hbar\frac{\partial}{\partial x^2} - e_3 i\hbar\frac{\partial}{\partial x^3} \right)$$

where c is the velocity of light and the basis vectors for the 4D spacetime have been written in 'symbolic' notation (the e_j) for the momentum ignoring the 4D equivalent of the 3D σ_j matrices used in Chapter 6. What was said about the non-relativistic momentum 'equations' is equally true about their relativistic analogues; they do not express any dynamical law. In particular, for a free particle they have nothing to say about the *relationship* between energy and the momentum components. The clue is in the non-relativistic momentum "equation" (i.e. identity). When one solves the momentum (or angular momentum) equation

$$-i\hbar\frac{\partial}{\partial x^j}\phi_k(x^j) = k\phi_k(x^j)$$

one gets positive and negative values of k in pairs. This does not mean that the *physical* component of momentum can be positive or negative, it simply means that the component can be numerically equal in each of two opposite *directions* in space. For the purposes of the interpretation of the values of momentum it might be better to write the (un-normalised) ϕ_k as

$$\phi_k(x^j) \stackrel{\text{def}}{=} \exp k(\pm|x^j|) \quad (\text{for } k > 0) \qquad (7.3)$$

to emphasise the point that there cannot be negative momentum, only positive momentum in different directions.

Now, when we deal with the relativistic energy/momentum vector given above, exactly the same thing occurs. In SM, the *energy* operator is

$$i\hbar\frac{\partial}{\partial t}$$

and the corresponding SM energy *identity* is, of course,

$$i\hbar\frac{\partial}{\partial t}\chi_\ell(t) = e_\ell\chi_\ell(t) \qquad (7.4)$$

with exactly the same solutions: positive and negative values of the *energy* e_ℓ in pairs. But this definitely does not mean that the (kinetic) energy of a free particle can be negative any more than the momentum of a particle can be negative but, exactly as in the momentum case, what it does mean is that there are two possible energies of the same magnitude in opposite directions of the relevant coordinate: in this case *time*. This is surprising, but this result is not a paradox because all particle mechanics of whatever flavour, since

Newton, give results which are reversible in time. In SM, the fact that Hamiltonian and the energy operators are scalars is so familiar that it is easy to see why confusion might arise when interpreting the eigenvalues of a vector operator.

In attempting to unify SR and SE, the role of another, much older, branch of physical science, statistical thermodynamics, is being ignored. The classical problem of reconciling a basic law of motion which is symmetrical in the direction of time, with the observed fact that the real world obstinately moves in one time direction only has been solved in the 19th century. The direction of time is not a property of the motion of individual particles but is determined by the motions of trillions of particles *via* statistical considerations. It is of the utmost importance to stress that, although the energy operator given above as part of the energy/momentum 4-vector, when used to construct an eigenvalue equation, gives rise to a eigenfunction equation which is an identity in the same sense as those associated with the momentum component operators. These equations generate those probability distributions in space and time which have constant momenta or energy and their formulation contains no involvement of any law of nature; in other words, they do not arise from the solution of an SE-like *equation*. And, of course, in all these equations the 'eigenvalues' k and e are continuous.

However, when a substantially identical *equation* of the form (7.4) arises, for example, from the separation of an SE into two, three or four equations each in a smaller number of space and/or time coordinates, the situation is quite different. In this case the law of nature *is* included and the separation parameter for energy (E, say) has both its sign and value *fixed* by the separated spatial (time-independent) part of the SE; E is a *separation parameter* of the original SE and must be used in place of e in (7.4).[6] Of course, upon the imposition of *constraints* on the motion, either by limiting the scope of the motion by confining the particle to a fixed, finite, region of space or, equivalently using cyclic boundary conditions, both of which are equivalent to the presence of a potential field, quantisation does occur on both momentum components and energy independently of any SE. It is easy to visualise confinement in space (by suitable 'walls' or cyclic motion) but it is not easy to see how one could impose a finite or cyclic constraint on motion in time.

What is now required is some attempt to find the *relationship* between the momenta and the energy which can be used to form the basis of a relativistic replacement for the non-relativistic SM.

7.3 A Relativistic Equation?

It was emphasised at the start of this chapter, on page 133, that Special Relativity (SR) does not involve a new law of dynamics but we can expect that there will be some more general scientific rules which apply in SR, in Classical Particle Mechanics (CPM) and in SM. In the absence of a new dynamical SR law it is necessary to find some equation which links the momenta and the (kinetic) energy since the historical way to obtain quantum mechanical equations is *via* such links. The conservation of energy is such a rule and is the best bet for establishing an equation linking momentum to energy in a relativistic scheme. In CPM, the momentum fixes the kinetic energy of a particle of mass m (T, say) since

$$E = T = \frac{|p|^2}{2m}$$

the question is: is there a relativistic equivalent which links $|P|^2$ to the kinetic energy of a 'relativistic particle'? In fact the corresponding equation linking P, when due allowance is made for the variation of mass and momentum *via* the Lorentz transformation, involves the squares of both the energy and the momentum:

$$P \cdot P = -\frac{E^2}{c^2} + p \cdot p = m_0^2 c^2 \quad \text{i.e. } E^2 = p \cdot p + m_0^2 c^2 \qquad (7.5)$$

which includes Einstein's famous formula for the energy a stationary particle ($E_0 = m_0 c^2$) which provides the intrinsic energy of mass m_0, the remaining contribution to the total energy (E). By making the now routine replacement of each component of the energy/momentum vector by the familiar HJE method, this will provide a set of operators and therefore a way to develop a

'Schrödinger-like' equation which satisfies the requirements of SR:

$$\hat{p}^j = -i\hbar \frac{\partial}{\partial q^j}$$

and E by

$$\hat{E} = i\hbar \frac{\partial}{\partial t}$$

But, awkwardly, *each* operator appears squared in the equation which leads to difficulties. However, it is clear that the LHS of Equation (7.5) is expressed as $|\boldsymbol{P}|^2$, the square of the modulus of $\boldsymbol{P}$ but what is needed is the *operator* $\boldsymbol{P}$ not its modulus. It has been shown in Chapter 6 that, in these circumstances, using an explicit (Geometric Algebra) notation in place of the usual formal representation of vectors gives a more realistic approach to the problem of the products and squares and square roots of vectors.

The 'symbolic' notation for the 4D basis vectors $\boldsymbol{e}_j$ can now be replaced by an explicit mathematical representation[7] as follows:

$$e_0 = \gamma_0 = \begin{pmatrix} \boldsymbol{I}_2 & \boldsymbol{0}_2 \\ \boldsymbol{0}_2 & -\boldsymbol{I}_2 \end{pmatrix} \quad e_j = \gamma_j = \begin{pmatrix} \boldsymbol{0}_2 & \boldsymbol{\sigma}_j \\ \boldsymbol{\sigma}_j & \boldsymbol{0}_2 \end{pmatrix} \quad \boldsymbol{I}_4 = \begin{pmatrix} \boldsymbol{I}_2 & \boldsymbol{0}_2 \\ \boldsymbol{0}_2 & \boldsymbol{I}_2 \end{pmatrix}$$

$$(7.6)$$

where $\boldsymbol{I}_n$ and $\boldsymbol{0}_n$ are the n-dimensional unit and zero matrices, respectively, and the $\boldsymbol{\sigma}_j$ are the 3D unit vectors of Chapter 6 and the operator $\hat{\boldsymbol{P}}$ becomes

$$\hat{\boldsymbol{P}} = i\hbar \left(\gamma_0 \frac{\partial}{c\partial t} - \gamma_1 \frac{\partial}{\partial x^1} - \gamma_2 \frac{\partial}{\partial x^2} - \gamma_3 \frac{\partial}{\partial x^3} \right) \qquad (7.7)$$

which may be substituted into Equation (7.5) to show that $|\hat{\boldsymbol{P}}|^2$ does indeed produce the required relativistic SE

$$\hat{\boldsymbol{P}}\psi = \boldsymbol{I}_4 m_0 c\psi \quad \text{or, in full:} \tag{7.8}$$

$$i\hbar \left(\gamma_0 \frac{\partial}{c\partial t} - \gamma_1 \frac{\partial}{\partial x^1} - \gamma_2 \frac{\partial}{\partial x^2} - \gamma_3 \frac{\partial}{\partial x^3} \right) \psi = \boldsymbol{I}_4 m_0 c\psi \quad (7.9)$$

where, of necessity, the solutions of the equation, ψ, must have a 4-component form

$$\begin{pmatrix} \psi_1 \\ \psi_2 \\ \psi_3 \\ \psi_4 \end{pmatrix} \tag{7.10}$$

each of which is, of course, a function of the x^j and t.

7.3.1   The Dirac Equation

Dirac arrived at a similar equation in his characteristic way of 'mathematics first, interpretation last' using quite a different approach; insisting that the correct relativistic equation should satisfy the equivalence of time and space. That is, not by squaring the time component of the energy/momentum as Schrödinger had done in his attempt to obtain a relativistic equation, but by insisting that the resulting equation be the same order in both the vector momentum operator components and the energy operator. His method also showed that this involved getting a square root of the relativistic operator and it resulted in his rediscovery of the geometric algebra basis vectors for a 4D space (spacetime) — now often called, in a quantum context, the Dirac matrices — which have similar properties to the γ_j matrices introduced above. It is certainly worth looking in more detail at Dirac's equation and both how its derivation differs from the above approach and, *a fortiori*, its interpretations.

Dirac took the relativistic expression for the total energy of a free moving particle

$$E^2 = c^2 \boldsymbol{p}^2 + m_0^2 c^4$$

and, as above, used the standard Cartesian replacement of the p_j and also sought a quantum mechanical equivalent of

$$E = \sqrt{c^2 |\boldsymbol{p}|^2 + m_0^2 c^4}$$

writing the required equation as

$$\hat{H}_D = -\left\{ i\hbar \left(c\alpha_1 \frac{\partial}{\partial x^1} + c\alpha_2 \frac{\partial}{\partial x^2} + c\alpha_3 \frac{\partial}{\partial x^3} \right) + \beta m_0 c^2 \right\} \psi$$

$$= i\hbar \left(\frac{\partial}{\partial t} \right) \psi \tag{7.11}$$

where the α_i and β were to be found from the condition that the Dirac Hamiltonian expression in brackets, $\hat{H}_D$, must, when squared, give the correct relativistic expression for E^2 above.

These conditions are, of course, the analogue in 4D space of the conditions quoted in the previous section which ensured that, in 3D space,

$$p^2 = |\boldsymbol{p}|^2$$

namely that, as above:

$$\alpha_i^2 = 1, \quad \beta^2 = 1, \quad \alpha_i\beta + \beta\alpha_i = 0$$
$$\alpha_i\beta + \beta\alpha_i = 0 \quad i = 1,3, \quad \alpha_i\alpha_j + \alpha_j\alpha_i = 0 \quad i \neq j \tag{7.12}$$

These matrices are, again, a form of the basis vectors of 4D spacetime and the fact that they are 4×4 matrices for a 4D space seems to bear only a coincidental relationship to the fact that the space is 4D since the corresponding matrices for 3D space are, of course, 2×2 not 3×3 as we discovered above.[8] In fact, one can easily show that when the above combination rules are represented by the simplest non-commuting entities — matrices — they must be of even dimension and at least 4×4.

Now, Dirac interpreted his expression as a Hamiltonian and, as such, should be used as the relativistic replacement for the SE since that had been his plan, to find an equation which replaces the SE when the constraints of SR are applied, that is an equation of the *form*:

$$\hat{H}_D \psi = \hat{E} \psi$$

replacing Schrödinger's $\hat{H}$ with his $\hat{H}_D$. His equation, therefore, kept the *scalar* energy operator $-\hbar\partial/\partial t$ on the right-hand-side of the equation.

This has the unfortunate effect of destroying the 4D-vector property of the energy/momentum vector by unwittingly using the scalar invariant (a constant) m_0c^2 as the time component of this 4D 'vector'. Dirac had succeeded in his quest of obtaining a SE-like equation which had the momentum components and the energy as first-order derivatives but at the cost of having a 'Hamiltonian' which is not obviously covariant. Fortunately, by multiplying[9] Equation (7.11) by $i\beta$, Dirac's equation is easily converted into the covariant form, (7.8), which also transforms the (α, β) matrix form to the γ_j form, making m_0c^2 into a scalar ($\beta^2 = 1$), and putting the energy operator into its proper place as part of the energy/momentum vector. However, Dirac's *interpretation* of his relativistic equation was, to say the least, unfortunate since it was based on his equation having a scalar energy operator and this interpretation has been carried forward to that of the covariant form.

7.4 Interpretation

Even before thinking about the interpretation of the relativistic SE it is worth saying just what has been achieved. When Equations (7.8) and (7.11) are solved they generate the allowed energy and spacetime distribution of any single particle of rest mass m_0 in unconstrained free motion. There is no mention, as yet, of any properties other than mass; the particle could carry a charge or not. Also the motion is not subject to any constraint, no limitation to the space in which it moves or, what amounts to the same thing, no potential field. This indicates that, strictly, unless such constraints are added, the system will not be quantised since neither of the operators are Hermitian in the field of their eigenfunctions; they are not capable of normalisation. It is now the rule to ignore these matters and treat the continuous values as the quantised energies of the system and, of course, to use the theory to describe a single charged electron which is perfectly acceptable, there being no potential field. With this in mind, the interpretations can be discussed.

Eigenvalues: First, and most problematic, the eigenvalues. The fact that the solution of the equations gives positive and negative energies for an isolated free particle having only kinetic energy led to a scramble to explain this surprise; the outcome is now very familiar.

Dirac's own initial response was to say that all the infinite lower-lying negative energy states must be full; thus any additional particles would have to fit into the states of positive (kinetic) energy. Even though there is no mention of *charge* in (7.11), he would have had in mind a relativistic equation for the electron. But, even then, that would mean that the whole system considered contained an infinite mass of electrons in which each particle repels all the others.[10] This would mean, at the very least, that all the electrons in these states would not have the energy of the eigenvalues of (7.11) but, in view of the enormous inter-electron repulsions, would require a self-consistent calculation of a different infinity. As regards the associated infinite mass density, surely we would be aware that we are moving about in a region of infinite gravity with the density of a neutron star. Clearly this interpretation is unacceptable, one might say in quite mild terms that this interpretation has 'no experimental support'.

An indirectly spectacularly successful development of this inter-pretation was the idea that an electron could be excited from one of the states of negative energy into a state with positive energy and the resultant 'hole' in the negative 'sea' would behave as a positive 'particle'. For example, in an electromagnetic field the sea of electrons would all move in one direction leaving the hole to apparently move in the opposite direction, simulating the behaviour of a positively-charged particle. At first, this was thought to be a proton but this is clearly not the case so a hitherto unsuspected particle must have the same mass as the electron but be positively charged: the positron. Such a particle was subsequently experimentally detected giving the 'sea and hole' interpretation a boost that has not yet quite disappeared in spite of the fact that Equation (7.11) contains no reference to charge. So, *post hoc* are we to assume that by sheer luck we have hit upon a uniquely-specialised Schrödinger-like equation, one which is only relevant to negatively charged particles? If so, there must, somewhere, be a relativistic 'master equation' which reduces to an equation for each type of particle when developed.

Much more interestingly, Feynman,[11] suggested that the negative energy states were electrons travelling backwards in time implicitly giving the correct interpretation: the fact that the energy operator is part of the energy/momentum *vector* as emphasised in Section 7.3 above when forming Equation (7.8). This is the only sensible interpretation and, for reasons outlined around Equation (7.3), it would be better for purposes of *interpretation* to give the solutions of the time factor in the solutions as

$$\chi_k(t) \stackrel{\text{def}}{=} \exp\left(ie_k|t|\right) \quad (\text{for } e_k > 0)$$

emphasising that the negative sign refers to the *direction* in which the particle moves in a given coordinate rather than the *value* of the energy which is always positive. Although now forgotten because of its association with Wheeler's[12] rather fanciful development, this interpretation is correct. So, rather than using an interpretation which allows time-travelling electrons, this interpretation gives a well-justified reason why the negative eigenvalues can be simply discarded as spurious since, unlike the spatial momentum components, where motion in both directions of the q^j, time only 'flows' in one direction.[13]

It is quite clear why Dirac was forced to make his exotic interpretations was due to the fact that, in his Equation (7.11), he had the energy operator as a *scalar* not a component of a spacetime *vector* ensuring that any negative eigenvalues must be interpreted as genuine if the theory is to be acceptable.

Eigenvectors: The solution of the Dirac and the covariant forms of the equation present no difficulty, since they are met anywhere the probability distribution is constant[14] and must, therefore, be of the form

$$\exp\left(i\lambda x^j\right) \quad |\exp\left(i\lambda x^j\right)|^2 = 1$$

where λ is any eigenvalue of one of the 4D coordinates x^j: (q^1, q^2, q^3, t). There are four equations of this simple form which turn out to be four simultaneous equations. The question then is: what does the existence of the four solutions mean?

It was seen in Chapter 6 that the introduction of the explicit form for 3D basis vectors (the σ_j) together with the presence in the system of an electromagnetic field (ϕ, A) led, quite naturally, to a term in the Schrödinger Hamiltonian which was interpreted as the interaction between an inherent quantity, 'electron spin', and the spatial components (A) of the field. It seems quite natural then to assume that the 4D form of the relativistic Equation (7.8) should, when formulated to contain the electromagnetic field, have some role related to the intrinsic properties of the particle if it carries an electric charge. That is, electron spin arises automatically when a multi-component state function is introduced either in 3D space or 4D spacetime which naturally poses the question — which will not be investigated fully here — does the same thing apply to an uncharged particle? Of course it does, the neutron has a magnetic moment but carries no charge. Notice that, in the derivation in Chapter 6, the spin-field interaction had a factor $\hbar e/2mc$ which contains the charge of the particle e so the theory outlined in that chapter would not allow a neutron to have an interaction with the components of the electric field vector ϕ or A. However, the two most familiar baryons, protons and neutrons, are not now regarded as 'fundamental' particles being composites of quarks. Even so, it does seem very strange that equations derived to describe the energies and probability distributions of particles whose only explicitly given properties are mass and, optionally, charge can detect whether or not a particle is composite.

There is another, perhaps more significant, matter associated with the introduction of an electromagnetic field; if a particle is charged or has a magnetic moment it will be accelerated by the interactions. This means that the assumption of uniform linear motion of the particle is no longer true and the assumption of a single velocity in the component parts of the Lorentz transformation are no longer valid; the Lorentz transformation matrix becomes a function of space and time as the particle's velocity varies. This matter was raised at the very start of this chapter and has been ignored in the rest of the chapter.

In many accounts of the interpretation of the Dirac Equation the first example used is that of a stationary electron. But there is no such thing as a stationary electron; an unstated axiom in all classical and quantum *dynamics* is that the particles are in motion. There is no HJE, no SE and no Dirac Equation for a stationary particle since all these equations depend on non-zero momenta for their formulation. Also, because of the Uncertainty Principle, a stationary particle has $\sigma q^j = 0$ and $\sigma p_j = 0$ whose product does not come close to exceeding $\hbar$.

In summary, the Dirac equation, particularly in its covariant form, *is* capable of an interpretation along the lines of the SE as outlined in the multi-component form of Chapter 6 which enables the exotic 'interpretations' so far offered by Dirac[15] and others to be discarded. If one can keep in mind that quantum theories are in a continuous development of an understanding of the real world and the fact that all these theories are based on what came before and that the interpretations must always be subject to saying something which can be, in practice or in principle, verified by observation and experiment, then the Dirac equation is a success. It has proved possible to obtain an equation obviously related to the Schrödinger Equation, but it is rather disappointing that it does not seem possible to obtain a *dynamical* law which could be substituted for the use of the static conservation of energy rule.

Endnotes

[1] It is not out of place to remark here that, in all the developments which stem from the formulations of CPM by Lagrange, Hamilton and Jacobi and the development of SM by Schrödinger, the underlying mechanical law is also that of Newton ('$F = \dot{p}$'). Neither of Lagrange's equation and the dynamical half of Hamilton's canonical scheme contain a new dynamical law, they are generalisations of Newton's insight, and Schrödinger's transition to QM is ultimately based on the same physical law but not for individual trajectories.

[2] For example, the index of Victor Arnold's famous book *Mathematical Mehods of Classical Mechanics*, Springer 1989, does not mention Special Relativity; a tacit acknowledgment of the fact that the mainstream of Classical Particle Mechanics is broken by Special Relativity.

[3] In this approach, I follow the long-forgotten work of R. C. Tolman, (*The Theory of the Relativity of Motion*, 1909) and M. Born, (*The Mechanics of the Atom*, 1927).

It is always useful to read the early attempts to get to grips with and interpret a new physical theory. These authors are struggling to *understand* the implications of the newer theory and its relation to older models before the new model becomes just another set of axioms to become familiar with and manipulate.

[4] *The Theory of the Relativity of Motion.*

[5] Which might be the proper time of SR.

[6] This is the source of yet another confusion about the 'energy operator' in QM; often the fact that the eigenvalues of

$$i\hbar \left(\frac{\partial}{\partial t} \right) \chi_j(t) = \ell_j \chi_j$$

are continuous is used to claim that $i\hbar \partial/\partial t$ cannot be the energy operator. But the eigenvalues ℓ_j cannot be chosen at random, they are fixed by the eigenvalues of other factors in the Schrödinger Equation.

[7] Writing the γ_j in bold may cause offence to those used to the standard typeface for these quantities but this is consistent

here where all vector quantities are given the honour of being emboldened. The mass m_0 and the velocity of light c are retained in the equations since what relativists call *natural units* clash with what chemists call *atomic units*.

[8] Unlike the present author, Dirac had to have the conviction, tenacity and skill to *derive* these conditions rather than simply take them from the mathematical literature. He was unaware of the history of the development of geometric algebra (as was practically every other mathematical physicist at the time) and so interpreted his equation as a set of four simultaneous scalar *equations* not as a single *vector* equation since the energy operator $\partial/\partial t$ on the RHS of his expression is clearly a scalar and not a vector component. In other words, Dirac had concentrated on keeping $\partial/\partial t$ as the distinct scalar energy operator.

[9] It is worth noting here what is achieved by simply 'multiplying' the 4D base vectors together. Remember these matrices are just representations of the unit vectors which, in contrast to a more familiar notation of, say (e_0, e_1, e_2, e_3) 'simple products' are allowed operations while the products the e_j are meaningless unless a dot $(\cdot)$ or a cross $(\times)$ is between them. This is because the matrix representations are the Geometric Algebra base vectors, whose properties were briefly mentioned in Chapter 6. Mainly, the product rule is:

$$ab = a \cdot b + a \wedge b$$

which, in 3D where the vector product is defined, becomes

$$ab = a \cdot b + ia \times b$$

In the case in hand, some of the products of the matrix representations are unit 4×4 matrices; non-zero scalar products of matrices, and some produce permutations of the αs and β. Thus the use of the matrix representation shields this important distinction of the generality of the GMA method.

$$\alpha_j \alpha_j = \alpha_j \cdot \alpha_j + \alpha_j \wedge \alpha_j = I_4 + 0_4$$
$$\beta\beta = \beta \cdot \beta + \beta \wedge \beta = I_4 + 0_4$$

$$\beta \alpha_j = \beta \cdot \alpha_j + \beta \wedge \alpha_j = 0 - \alpha_j$$

$$\alpha_1 \alpha_2 = \alpha_1 \cdot \alpha_2 + \alpha_1 \wedge \alpha_2 = 0_4 + i\alpha_3$$

$$(7.13)$$

The fact that Dirac's equation can be converted to the covariant form so easily hinges on the ability to use the GMA full product which when multiplying by the matrix representation leads to both scalar and vector entities. This is impossible using the symbolic dot or cross product.

[10] Perhaps enough repulsive energy to start a new 'big bang'!

[11] Feynman, R. The theory of positrons, *Phys. Rev.* **76**(6) (1949), p. 749, following an idea Stueckelberg E. C. G. Stueckelberg, Relativistisch invariante Strungstheorie des Diracschen Elektrons, *Ann. Phys.* **21** (1934), pp. 367–389 and 744.

[12] Described in Feynman's Nobel Lecture of 1965.

[13] In another interesting example, involving the problem of the direction of time in physics, also involving Richard Feynman and John Wheeler, occurs in classical electrodynamics. When the Maxwell equations (which are covariant) are transformed into the form containing the potentials ϕ and A and solved, the solutions are symmetric in both directions in time. The solutions are called the retarded potentials (time forward) and advanced (time backwards) solutions. Although it is obvious that, like any form of mechanics, this is due to the time-symmetry of the Maxwell equations this has been picked up by amateur philosophers to generate more exotic theories particularly since the apparent paradox carries over into quantum electrodynamics, where in the case of physical interpretation, anything goes.

[14] Notice that, if V (the 'volume' in which the particle may move) is given by

$$\int_a^b |\exp{(i\lambda x^j)}|^2 dx^j = V_{ab}$$

then, unless the motion is constrained to move in a finite range a, b then

$$\int_a^b |\exp{(i\lambda x^j)}|^2 dx^j = V_{ab} \to \infty$$

[15] This chapter has been very unkind to Paul Adrien Maurice Dirac but only, as in the Preface, unkind to his *interpretations* of his astonishing bold development of quantum theories. If he had not obtained his theory of the chemical bond and, above all, the subject matter of this chapter, the scientific world would have been a good deal poorer.

Chapter 8

Fields and Second Quantisation

I prefer to doubt rather than define what is unknown

John of Salisbury (1115–1180)

8.1 Fields and Particles

Most introductions to a quantum theory of fields will mention the wave–particle duality and say something along the lines that, since the electromagnetic field has both continuous and particle manifestations, it will not be out of place to investigate the question of whether or not there are fields underlying the behaviour of all particles. In this approach the most natural thing to assume, it is said, is that the state functions of particle QM play an analogous role to the EM field and the particles are just the 'quanta' of this field in the same way as the photons are said to be quanta of the EM field.

There is a rather large lacuna[a] in this reasoning:

> The EM field is a real massless substance existing 'out there' in the physical world while the state function ψ (when squared) is a probability distribution; a mathematical function which exists in our heads, on paper or in our computers. *If their were no humans ψ would not exist, but the EM field would still be there.*

What then will be the quanta associated with the field of probability? Will they also be concepts in our heads or will they be ontologically

---

[a]More graphically in the current idiom an 'Elephant in the room'.

real? If they are to be real how did they come to be real from a purely mathematical 'field'?

I should say at this point that I am suspicious of all arguments based on perceived qualitative analogy or claimed 'symmetry' alone and particularly of this attempt at the unification of the interpretation of two very different fields. In fact, there is an 'anti-symmetry' argument here which is at least as strong:

In the SM case, the particles are real and the probability distribution is a concept while in the EM case, the field is real and the particles — photons — are conceptual.

Comparing the familiar particles (electrons, protons, mesons, etc.) and the EM field, there is an obvious asymmetry: Individual types of particle are the same as all the others — all electrons are the same — but there are as many different photons as there are frequencies of vibration in the field; an uncountable infinity.

This distinction cannot be decided by its formal convenience or by ratiocination — by thinking very hard about it — we must appeal to experiment to try to find a way forward. It might be a useful start to 'define' a particle as an entity which can only be in *one place* at a given time and a field as something which is *everywhere* in the space considered at the same time.

On the question of the existence of the referents of the formalism, one thing that must be said before investigating the quantum theory of fields is to comment on the way in which any experimental result may be interpreted and, if possible, verified. I do not want to imply any regression to 'instrumentalism' or any form of positivism, but this reference to the instruments of measurement must be stressed as it seems to have been overlooked for nearly a century.[1]

8.2   Experiments — Particles or Waves?

All measurements of the energies or momenta etc. of molecular, atomic or subatomic properties are made with *material* instruments whether these measurements are suspected (or known) to be of the properties of waves or particles. From the clicks on the Geiger counter, the photo-electron effect or diffraction fringes, the presence, tracks or properties of entities result from the incident species, exciting in some way one or more properties of the material of which the detecting device is constructed. What one 'sees' is not the entity or one of its properties but the result of this interaction.

For example, the flashes on a phosphorescent screen which indicate the build-up of a diffraction pattern are not the incident photons, electrons or radiation but the reaction of the material in the screen which has absorbed the impulse energy or momentum. It is not possible from this experiment alone to derive any conclusion about the nature of the entity causing the flashes.

Now, it is obvious that, at the scale of these interactions, the energies, momenta etc. of all materials are *quantised*. Thus, for any detection, the incident object must provide one of the allowed excitations of the detecter material. This, of course, may explain the existence of a threshold below which detection is not possible but what it does not and cannot mean is that the exciter *itself* had quantised levels. One sees discrete properties, not because the incident property is quantised (although it may be, of course) but because the material of the detector *can only report the excitation of its own allowed transitions* because of the sub-atomic scale of the properties involved. Thus, one can only say that the incident entity, among its possibly numerous or even continuous properties, had one which matched the quantised energy gap of the detector material. As Einstein is reported to have remarked:

> Just because beer comes in pint bottles does not mean that it exists everywhere and at all time in pints.

The upshot of these considerations is that these kinds of measurement techniques cannot distinguish between field oscillations ('waves') and particles as has been shown originally by Duane for diffraction almost a century ago and for most, if not all of what are still thought to be the archetypal and decisive evidence for the existence of photons.[2] Equally important, one cannot tell from this type of experiment whether or not the incident property is quantised. Of course, the photo-electron effect tells us that the energy of EM radiation depends crucially on its frequency not simply on its amplitude but not that this energy is quantised.

So, we see electron microscopy, photo-electric effect, Compton scattering and electron diffraction but we cannot tell (from this source alone) what is causing these effects, whether 'particles' or 'waves'. The crucial evidence for this distinction has to come from other experiments, for example deflection by electromagnetic fields. Or, as has been the practice for some time, simply ignore the distinction by giving it a name: the wave–particle duality.

The very existence of the wave–particle duality fantasy is due to the failure to see that it is, at the sub-atomic level, the properties of the excitations of the measurer we see, not the properties of the exciter directly. The now-famous experiment of the diffraction of a beam of electrons, which is the 'definitive proof' of the duality, assumes that the diffraction 'screen' is a simple homogeneous material without any atomic or sub-atomic structure and simply exists to provide a pattern of slits whose sole function is to create a diffraction pattern from the incident beam. But at this scale of phenomena *all* matter is composed of nuclei and electrons, and any electrons confined to the very narrow bands of material between the slits are, of course quantised in momentum and energy. If the screen is of a metallic nature, the electrons in the narrow inter-slit bands can be assumed to be the least tightly bound 'conduction electrons' and have momenta of approximately integral multiples of $\hbar/\ell$ where ℓ is the width of the narrow bands. These electrons will collide with some of the electrons in the beam and deflect them in a direction perpendicular to the beam's direction and, as the detailed mathematics shows, produce the experimental 'diffraction pattern'. *In other words, it is the fact that electrons in the material of the screen are quantised and are creating the pattern not the 'diffraction' of the unquantised beam.*

8.3   The Lagrangian Method in Field Theories

The quantum theory of continuous substances, like Schrödinger's Mechanics (SM), has to be preceded by the classical theory of such systems but, as we have seen in Chapter 7, historically the necessity of making such a theory relativistically covariant is said to have excluded Schrödinger's original *method* since it is obvious that the variational expression he used is not Lorentz covariant.

The most common approach to Quantum Field Theory (QFT) is to say:

- Field theories must be relativistic, i.e. Lorentz covariant.
- Since Hamilton's canonical equations for particle theories are clearly not Lorentz covariant — they give a special role to time — it is preferable to start from the Lagrangian approach.

If, in exactly the same way that Schrödinger 'derived' SM, the Lagrangian is written in terms of a field Lagrangian density:

$$L() \to \mathcal{L}()$$

then Hamilton's principle becomes

$$\delta S = \delta \int L() dt = \delta \int \mathcal{L}() dV' dt = 0$$

where dV' is the volume element for all the space coordinates of the system. Simply by using Minkovski's metaphorical 4D spacetime, we then have

$$\delta S = \delta \int L() dt = \delta \int \mathcal{L}() dV = 0$$

where dV is the spacetime 'volume' element. This form of Hamilton's principle is then said to be 'manifestly Lorentz covariant' because the combined space and time coordinates mean that the spacetime volume element is a scalar and, *provided that $\mathcal{L}$ is a scalar*, the variational problem presented is covariant.

But these arguments are misleading since Schrödinger's *method* is not based on the Hamiltonian of the classical canonical equations for individual particle trajectories but on the Hamilton–Jacobi Equation (HJE) for all possible trajectories. When one looks at the QFT literature, it is immediately apparent what has happened. The word

'Lagrangian' has two meanings in current usage:

(1) To physicists, it originally meant a quantity in the mathematically-articulated science of (particle) mechanics. It is given in terms of configuration-space coordinates and the time derivatives of these coordinates — velocities — and, in general, time: in particular:

$$L(q^j, \dot{q}^j, t) = T - V$$

where T is the kinetic energy (function of q^j and $\dot{q}^j$ usually) and V is the potential energy (usually a function of q^j and occasionally t). Hamilton's principle,

$$\delta S = \delta \int L()dt = \delta \int \mathcal{L}()dV = 0$$

generates the Lagrange equations of dynamics and thence onward to Hamilton's canonical equations and the HJE.

(2) To mathematicians, only the variational principle is retained, the physical interpretation in the terms of mechanics is abstracted away. In this context, the term 'Lagrangian' simply means some functional which is to be extremised (minimised, usually) and the variables on which the functional depends may well have no particular physical referent.

Now, using the latter usage of the term Lagrangian, the original variational method of Schrödinger is a Lagrangian method since it is just the method of finding the extreme of

$$\int \{\rho_H - \rho_E\}dV$$

where both ρ_H and ρ_E are distributions in space and time and dV is — in Cartesians for simplicity — $dxdydzdt$.[3] Sure enough, when one looks at the Lagrangians used, for the meson field for example, they are identical to the ones which Schrödinger would have used from his considerations of the HJE and which, if solved in the same way, would give an SE for mesons, identical to the electron equation except for particle mass and charge. Strictly speaking, the H in ρ_H above is not the correct notation since it is the 'momentum terms' in the HJE not the explicit momenta of the classical Hamiltonian which appear in Schrödinger's functional. In short, Schrödinger' method *is* a Lagrangian method, does not start from Hamilton's canonical equations and it is the particular Lagrangian used, not the method itself, which makes the *overall approach* non-relativistic.

8.4   Two Approaches to Schrödinger's Functional

From what has been said in the last section it is obvious that, in order to obtain the quantum equations for fields, it is not necessary to look any further than Schrödinger's method in Chapter 2 since Schrödinger's method is already a method for determining fields — at least mathematical ones — the function $\psi(q^j)$ which are solutions of the SE, are, when squared, fields of probability of presence in space. However, the two possible ways of looking at the mathematical Lagrangian method of Schrödinger's functional illustrates and contrasts the 'particle' and 'field' interpretations of the resulting solution of the variational problem, the one used by Schrödinger where the solutions are the state *functions* of space and time and the field approach where the solutions are defined to be *operators*. For the purposes of this work — the interpretation of SM — only the theory of the fields which are the solution of the Schrödinger Equation (SE) will be examined, precisely because the case of 'real' electromagnetic fields is outside the scope of SM.

Looking back, the Euler–Lagrange equation which generates these solutions for Schrödinger's functional was

$$\frac{\partial \mathcal{L}}{\partial \psi} - \sum_{j=1}^{3} \left\{ \frac{1}{\sqrt{g}} \frac{\partial}{\partial q^j} \left(\sqrt{g} \frac{\partial \mathcal{L}}{\partial (\frac{\partial \psi}{\partial q^j})} \right) \right\} - \frac{\partial}{\partial t} \left(\frac{\partial \mathcal{L}}{\partial (\frac{\partial \psi}{\partial t})} \right) = 0 \quad (8.1)$$

and so the problem of finding the equation for a continuous field was solved already for a particular 'Lagrangian'. All that is necessary is to have the relevant Lagrangian for the problem in hand and the solution follows by solving the Euler–Lagrange equations generated from that particular Lagrangian. In many introductions to the theory fields this material is developed from the beginning and transformed into a different notation by replacing the spatial variables q^j by the *functions* ψ and ψ^*.

One defines the 'functional derivative' of the Lagrangian by the first two terms of the above equation

$$\frac{\delta \mathcal{L}}{\delta \psi} \stackrel{\text{def}}{=} \frac{\partial \mathcal{L}}{\partial \psi} - \sum_{j=1}^{3} \left\{ \frac{1}{\sqrt{g}} \frac{\partial}{\partial q^j} \left(\sqrt{g} \frac{\partial \mathcal{L}}{\partial \left(\frac{\partial \psi}{\partial q^j} \right)} \right) \right\} \tag{8.2}$$

which gives for the Euler–Lagrange equation

$$\frac{\delta \mathcal{L}}{\delta \psi} - \frac{\partial}{\partial t} \left(\frac{\partial \mathcal{L}}{\partial \left(\frac{\partial \psi}{\partial t} \right)} \right) = 0 \tag{8.3}$$

and, if we may risk use of the 'dot' notation for the *partial* derivative with respect to time, i.e. $\dot\psi$ the final compacted equations becomes

$$\frac{\delta \mathcal{L}}{\delta \psi} - \frac{\partial}{\partial t} \left(\frac{\delta \mathcal{L}}{\delta \dot\psi} \right) = 0 \tag{8.4}$$

which has the same structure as the CPM Euler–Lagrange equation for a particle trajectory

$$\frac{\partial L}{\partial q^j} - \frac{\partial}{\partial t} \left(\frac{\partial L}{\partial \dot q^j} \right) = 0 \tag{8.5}$$

The fact that, in the former of these two equations

$$\dot q^j \stackrel{\text{def}}{=} \frac{dq^j}{dt}$$

and, in the latter,

$$\dot\psi \stackrel{\text{def}}{=} \frac{\partial \psi}{\partial t}$$

makes this analogy somewhat strained but it has proved tempting and has been used as a justification for the method of Second Quantisation.

The technique is to obtain the field equivalent of the momentum of a particle by replacing:

$$p_j \overset{\text{def}}{=} \frac{\partial L}{\partial \dot{q}^j} \quad \text{by} \quad \pi_j \overset{\text{def}}{=} \frac{\delta \mathcal{L}}{\delta \dot{\psi}}$$

and then seeing what is necessary to make the results satisfy the analogue of the 'particle quantum condition' for π and ψ, i.e. something like

$$(\psi\pi - \pi\psi) = i\hbar \tag{8.6}$$

Clearly this is not possible if ψ and π are scalar functions since their product always commutes. Both ψ and π must be *defined* so that they can be interpreted as operators in some sense and, of course, ultimately, operators must have some functions, or vectors in the general sense, to operate on. It is not clear if or how this can be done since there is no indication of the form of the operators $\hat{\pi}$ and $\hat{\psi}$ or, crucially, for such a relationship to be shown, what the operators operate on. The dependence of the operators on the original spatial coordinates and time is assumed to be

$$\hat{\psi}_j \overset{\text{def}}{=} \hat{\psi}(q^j, t_j) \quad \hat{\pi}_j \overset{\text{def}}{=} \hat{\pi}(q^j, t_j)$$

and, by analogy with the theorem of SM, the relationship between the operators is *defined*[4] to be that given in Equation (8.6) for each point in space and time with the standard notation

$$\begin{aligned}
[\hat{\psi}_j, \hat{\pi}_j] &= (\hat{\psi}_j \hat{\pi}_j - \hat{\pi}_j \hat{\psi}_j) = i\hbar \\
[\hat{\psi}_j, \hat{\psi}_k] &= 0 \\
[\hat{\pi}_j, \hat{\pi}_k] &= 0 \\
[\hat{\psi}_j, \hat{\pi}_k] &= 0
\end{aligned} \tag{8.7}$$

without the inconvenience of saying what the operators operate on. This suggests that a continuous field is treated as being, in some way, equivalent to an infinite set of particles, one for each point in space.

To investigate this approach, it is useful to work with a specific example, the single particle Schrödinger Equation, to see how it works in practice. In this case the Lagrangian is given by Equation (8.1) and the two $\hat{\pi}$s conjugate to $\hat{\psi}$ and $\hat{\psi}^*$ are easy to evaluate since there is only a single occurrence of $\dot{\psi}$ and no occurrences of $\dot{\psi}^*$

$$\frac{\delta L}{\delta \dot{\psi}} = \pi = \psi^*; \quad \text{and} \quad \frac{\delta L}{\delta \dot{\psi}^*} = 0$$

This result seems opaque at the very least — how is it possible to interpret this 'field momentum' and why is there no field momentum conjugate to the 'field coordinate' ψ? Some slight help may come from a consideration of the Lagrangian associated with a time-independent Lagrangian where in Equation (8.1)

$$\psi^* \frac{\delta}{\delta t} \psi \rightarrow \psi^* E \psi$$

E being a scalar energy. In this case, both the possible πs are zero. In the particle probability of presence interpretation of ψ each of the trajectories in the set of all possible ones with energy E has a momentum of course, but the *distribution in space* of momentum does not change, that is, the particles are moving but the *field* of probabilities of presence does not 'move': it does not evolve in time. So, one might conclude that the field momentum is related to — a measure of — the time-evolution of the field. It looks as if the only possible interpretation of the field momenta of the Schrödinger Lagrangian is that ψ^* is not an independent variable in this case. Certainly, in the time-independent case the SE is real and it has real boundary conditions so the solutions may always be chosen to be real.[5]

While looking at the use of Schrödinger's Lagrangian method, it is worth recalling — yet again — the distinction between the mathematical technique and the dynamical law being invoked:

• The dynamical law in this case is the equality of the mean value of the Hamiltonian density and the energy density, what has been called the Schrödinger Condition.
• The Lagrangian method used to find that equation and its boundary conditions ensures that the dynamical law is obeyed.

All the use in physics of the Lagrange methodology is of this type, that is, the technique is used to ensure that:

Something = Something else
or, as it is usually formulated
Something − Something else = minimum, preferably zero.

The law of nature has to be contained in the functional[6] to be minimised; the Lagrangian method simply transforms this law into a more useful pair of equations or conditions. Most of the known laws of mechanics and EM theory can be cast in this form although this is, in the case of EM theory, *post hoc*. In these cases, the dynamical laws are already known, but in the case of 'fundamental' particle physics for which the field equations are sought the opposite is true; the laws of the interaction mechanisms of these particles — in the majority of cases — are not known. This can be made, by using experimentally determined parameters and use of the Lagrangian method, to give extremely accurate results. But this is going very far from SM where the laws must be known and the physical interpretation of the theory is paramount. It is, using a parametrised approach, quite often easy to obtain the correct numerical answers but this is not the main aim in science which is to understand what is happening and what is causing it to happen.

It is clear that what is necessary for the development of the functional approach of the theory of fields is that both

$$\frac{\delta L}{\delta \psi} \quad \text{and} \quad \frac{\delta L}{\delta \psi^*}$$

appear in the functional to be minimised, which is definitely not the case in SM. But, it must be kept in mind that these considerations are all simply *analogies* between the functional derivatives and ordinary partial derivatives and that their interpretation is problematic; can we really simply by *ex cathedra* pronouncement transform a function of 3D space ($\psi*$) into an operator which works on some, as yet unknown functions? What does this operator do to the state function ψ which solves SE in Equation (8.6)?

At this *impasse* it is usual to introduce the harmonic oscillator Hamiltonian and its eigenfunctions to provide another analogy. It is well known that the adjacent eigenvalues of this system are separated by the same energy gap, say ΔE, and that the Hamiltonian can be expressed as the product of a pair of 'ladder operators' since the momentum, $\hat{p}_x$ say, and the conjugate coordinate, x, both appear squared in the Hamiltonian. These operators, when operating on a state function of the system, transform the state function into an adjacent one of either increased or decreased energy by an amount ΔE. The operators can then generate the whole set of solutions of the harmonic oscillator by supplying any integer multiple of ΔE. This system, then is an analogy for operators which increase or decrease the number of *particles* associated with the 'field' ψ, say. But what the operators actually do is to mimic the addition or subtraction of energy from the system which also changes the probability of presence for the position of the oscillator; there are no particles involved. If there were particles generated ('phonons'?), where would the equations of motion for these particles come from?[7]

8.5   Interpretation

It is quite clear that the interpretation of these theories of quantising the mathematical fields generated by SM, which is the purpose of this work, is going to be quite challenging. The analogy between exciting an oscillator and creating a particle is, if anything, disturbingly revealing rather than helpful. The SE for a simplified 1D harmonic oscillator is:

$$\hat{H}\psi_n = \left(\frac{1}{2}k^2x^2 - \frac{\hat{p}^2}{2m}\right)\psi_n$$

$$= \left(\frac{1}{2}k^2x^2 - \frac{\hbar^2}{2m}\frac{d^2}{dx^2}\right)\psi_n = E_n\psi_n \qquad (8.8)$$

where k the (scaled) force constant and the momentum operator $\hat{p}$ is, as usual

$$\hat{p} = -i\hbar\frac{d}{dx}$$

This Hamiltonian, $\hat{H}$, may be written[8] as the product of the two mutually adjoint ladder operators:

$$\hat{\ell}_{up} = \frac{1}{\sqrt{2}} \left(kx - \frac{i}{m}\hat{p} \right)$$

$$\hat{\ell}_{down} = \frac{1}{\sqrt{2}} \left(kx + \frac{i}{m}\hat{p} \right)$$

(8.9)

and these two operators have a 'raising' and 'lowering' effect on the ψ_n:

$$\hat{\ell}_{up}\psi_n = \sqrt{n+1}\,\psi_{n+1}$$

$$\hat{\ell}_{down}\psi_n = \sqrt{n-1}\,\psi_{n-1}$$

But these operators have no physical interpretation; they are simply a mathematical convenience useful to generate the state functions of the harmonic oscillator; the actual physical process of exciting an oscillator from a given state to a higher one is an *experiment* involving the input of energy (radiation, say) to the system. To take a specific example, a vibrating diatomic molecule has to have a non-zero dipole[9] moment to be excited by infrared radiation and the allowed transitions can be calculated by perturbation theory but the ladder operators are nowhere to be seen in these calculations. What is very clear from these considerations, if it were not clear earlier, is that this analogy with the harmonic operator is not a help to the interpretation of quantum fields but is just a mathematical analogy for the 'interpretation' of manipulations of the state functions not a description of something which actually happens.

The operator $\hat{\psi}^j$ is sometimes interpreted as creating a particle at the point q^j, t_j in space within the field, but this is fraught with difficulties of interpretation particularly if the particles are mutually interacting. A more useful approach is to expand the field of a linear combination of functions spanning the same many-particle space as ψ^j along the lines of perturbation theories with the operator coefficients representing the addition or subtraction (redistribution) of particles among these 'state functions'. This step reveals what second quantisation really is: it is a *mathematical tool* for manipulating linear expansions of approximate state functions when the number of particles may not be constant.

This change of variable introduced by going to the apparently more fundamental choice of the functions ψ and ψ^* in place of the coordinates q^j may well have generated operators in addition to the functions ψ and ψ^*, but this would have to be shown rather than assumed on the basis of an analogy. Equally important, the interpretation of these operators and of the species on which they operated would have to be interpreted. In fact the justification of the method of second quantisation by analogy with the Lagrangian is not often given, and second quantisation is now usually introduced axiomatically for what it is: a computational technology for the manipulation of approximate state functions for many-particle systems.[10]

In short, the method of second quantisation falls short of the main requirement of SM: it is not physically interpretable. The terms in the expansions do not, in general, correspond to physical states or processes. Each step in the process of the second quantisation of the SE is an analogy, from the initial assumption that the state function is an ontologically real object, *via* the definition of the properties of the 'operators' $\hat{\psi}$ and $\hat{\pi}$, to the strange use of the properties of the harmonic operator ladder operators. Analogies are sometimes useful when thinking about new theories but they are just that: analogies not laws. The interpretation of SM has been dogged by the unthinking use of analogies from the very beginning, starting with the disastrous analogy between the mathematical form of the SE and the wave equation; it is time to minimise their use and treat them with care.

Endnotes

[1] Large areas of contemporary theoretical physics have now almost abandoned the irksome requirement of experimental verification; I am thinking of string theory and its variants and cosmology. The case of cosmology is the most robust and unashamed, simply inventing new entities in order that the theory be saved — dark matter, dark energy, dark flow — while the string theorists appeal, like Dirac of old, to the alleged beauty of the mathematics. In theoretical physics, Edward Thompson's soaring Eagles of Chapter 2 are getting the better of the Great Bustard.

[2] See *Zeitschrift für Physik* **41**, p. 828 by G. Wentzel (1927), for example, for the photo-electric effect. In fact, it is not going too far to say that the problem of *interpreting* QM has slept for nearly a century while the formalism stormed ahead, not wishing to be hindered by such niceties.

[3] A historically consistent name for Schrödinger's functional would, of course, be the 'Schrödingerian' but this is such a tongue-twister that it is, rather unfairly, unlikely to be adopted.

[4] Notice, once again, that this method is obliged to *define* the properties of mathematical objects in order that the subject matter can describe quantum effects. This is entirely foreign to SM since, in SM, the commutation properties of the coordinates and momentum operators arise naturally from their explicit mathematical representation. No additional definition is needed and, of course, this result is inherited from the HJE.

[5] In the case of degeneracy of E they may also be chosen as complex conjugate pairs of course.

[6] In fact, the whole 'scheme' of:

'Lagrangian Functional'

$\rightarrow$ 'Momenta' $\rightarrow$ 'Hamiltonian'

$\rightarrow$ 'Hamilton–Jacobi Equation'

is very general and flexible and can be applied to any system undergoing change, completely independent of any form of mechanics as emphasised in Chapter 7.

[7] Similarly, where are the equations of motion for the photons; particles of the EM field?

Incidentally, many quantum mechanical Hamiltonians separate into a product of ladder operators as originally shown by Infeld and Hull in 1951 (*Reviews Modern Physics* **23**, p. 21) and followed up by Coulson and Joseph in 1967 (*Reviews Modern Physics* **39**, p. 829). In such systems, it is common for the gaps between the levels to not to be equal, which rather spoils the analogy.

[8] Done by using $(a^2 - b^2) = (a - ib)(a + ib)$, notwithstanding the fact this simple formula depends on a and b commute, i.e. $ab = ba$. In fact, when $a \to x$ and $\to \hat{p}$ this is not the case since:

$$x\hat{p} - \hat{p}x = -i\hbar \left(x\frac{d}{dx} - x\frac{d}{dx} - \frac{d}{dx} \right) = -i\hbar \left(-\frac{d}{dx} \right)$$

[9] Strictly, a dipole or higher multipole but even the quadrupole case is much more difficult to detect; the difficulty in detecting homopolar diatomics in interstellar space is well known although atomic hydrogen is very easy to find. A retired chemist might think, 'Hmm, atomic hydrogen is very abundant in space, a hydrogen molecule is much more stable than two hydrogen atoms, the non-polar hydrogen molecule is difficult to detect, is this the source of dark matter?'

[10] This is not to belittle the usefulness of second quantisation methods, even in systems where the number of particles is constant, the method is a useful substitute for the manipulation of determinants of single-particle functions, it is easy to programme since there is no interpretation needed.

Epilogue

It may well seem to anyone reading this book that the author set out to give a hagiography of Schrödinger at the expense of the other great pioneers of QM. That is, there was an existing prejudice of Schrödinger's pre-eminence. The problem here is that it was precisely the opposite; unfortunately there is no word '*post*judice' in English which would mean the exact opposite of the word '*pre*judice', that is, the material is judged *after* having been carefully considered not before as implied in 'prejudice'. The fact is, the more an unprejudiced reader looks at those first papers of 1926, the more convinced does one become that Schrödinger had, from the point of view of an interpretable QM, almost 'said it all'. That is, Schrödinger needs no hagiography from me or anyone else, although the temptation is rather compelling since, in addition to his contributions to quantum theory, relativity and statistical mechanics, he started molecular biology in his spare time, as it were.

Although many of the 'classic' mysteries and paradoxes are solved by the considerations involved in this work, *real* difficulties of interpretation still remain. The most prominent one emphasised above is the meaning of the fact that the 'mean value' nature of the Schrödinger Condition necessarily involves that quantum Hamiltonian function is not always equal to the energy of a system. This genuine break with 'tradition' does not commonly appear in the various interpretations of quantum theories but an explanation is sorely needed.

Setting this aside, for myself I would like to see a resolution of the first paradox of even pre-Schrödinger quantum theories: the fact that the dozens of accelerated electrons in atoms and the hundreds in many molecules do not emit electromagnetic radiation which would cause matter to disappear in microseconds. Einstein also commented on the fact that there seem to be no gravitational waves arising from the very large percentage changes of mass predicted by Special Relativity in the case of the inner electrons of heavy atoms whose velocities are predicted to exceed the velocity of light in calculations based on (non-relativistic) Schrödinger's Mechanics (SM).

Finally, some opinions on the relationship between science and mathematics; one from Schrödinger and two of Dirac's:

"A theoretical science unaware that those of its constructs considered relevant and momentous are destined eventually to be framed in concepts and words that have a grip on the educated community and become part and parcel of the general world picture theoretical science, I say, where this is forgotten, and where the initiated continue musing to each other in terms that are, at best, understood by a small group of close fellow travellers, will necessarily be cut off from the rest of cultural mankind; in the long run it is bound to atrophy and ossify however virulently esoteric chat may continue within its joyfully isolated groups of experts."

E. Schrödinger, *Are There Quantum Jumps? British Journal for the Philosophy of Science* **3** (1952) p. 109.

"A good deal of my research work in physics has consisted in not setting out to solve some particular problems, but simply examining mathematical quantities of a kind that physicists use and trying to get them together in an interesting way regardless of any application that the work may have. It is simply a search for pretty mathematics. It may turn out later that the work does have an application. Then one has had good luck."

P. A. M. Dirac, Pretty Mathematics, *International Journal of Theoretical Physics* **21**(89) August (1982), p. 603.

"The interpretation of quantum mechanics has been dealt with by many authors, and I do not want to discuss it here. I want to deal with more fundamental things."

P. A. M. Dirac, *The Inadequacies of Quantum Field Theory*, in Paul Adrien Maurice Dirac, B. N. Kursunoglu and E. P. Wigner (Cambridge University, Cambridge, 1987) p. 194.

Schrödinger is right and Dirac is wrong; there *is* nothing more fundamental than the conceptual understanding of the structures and processes of the world; mathematics is a tool.

Suggested Reading

Obviously, the first item to look at is an English translation of Schrödinger's pioneering papers which have been constantly referred to in the text:

Wave Mechanics by Erwin Schrödinger, originally published in 1928 by Blackie, the translation was done by a lady who knew German but no Physics and a physicist who knew no German. The book has been extended to cover more of Schrödinger's work and published by Chelsea, 1982.

Here is a very short list of books which might refamiliarise the reader with the salient points of Classical Particle Mechanics and how it links to Schrödinger's Mechanics:

Analytical Mechanics for Relativity and Quantum Mechanics by Oliver Davis Johns, (second Edition) OUP 2006, very comprehensive and comprehensible — excellent.

The Variational Principles of Mechanics by Cornelius Lanczos, (fourth edition) Dover 1970. Beautiful book by a man who loved mechanics.

A History of Mechanics by René Dugas, Dover 1988. A good coverage of the development of mechanics from the Hellenistic period to the rise of Relativity and Quantum Mechanics. It is a little partisan for the contributions of the European scientists which goes some way to balance the anglophone emphasis on the work of Newton and Hamilton.

Variational Principles in Dynamics and Quantum Theory by Wolfgang Yourgrau and Stanley Mandelstam, (third edition) Dover 1968. This is interesting in its own right but especially for a postscript on page 125 which quotes a communication from Schrödinger pointing out that the use of

$$\hat{p}_j = -i\hbar \frac{\partial}{\partial q^j}$$

is only valid in Cartesian coordinates and the genuine expression for ∇^2 is the one given in Chapter 2. As usual, this has been ignored for decades.

The above books should be read simultaneously dipping into whichever one seems relative to a path of thought.

Philosophy and the Physicists by L. Susan Stebbing, first published during World War II, Dover reprint 1958. Here is a 'must read' with some plain talking. Lizzie Stebbings' book is 'out of date' but it is a clear polemic of the very best kind against idealist philosophy. All that is necessary is to mentally substitute the names of Jeans and Eddington by the more familiar contemporary purveyors of idealist philosophy of which there are more than just two! The book is long out of print and I am trying to produce an e-book version which should appear on the web in due course.

Index

Printed in the United States
By Bookmasters